大道至简

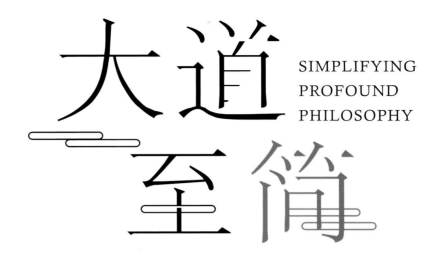

SIMPLIFYING PROFOUND PHILOSOPHY

台式空间设计品鉴
TAIWANESE SPACE DESIGN APPRECIATION

深圳视界文化传播有限公司　编

中国林业出版社
China Forestry Publishing House

PREFACE
前言

Living a new life in the 21st century, the socialist form we were accustomed to has gradually become weak while individual thought which replaced it increasingly becomes mature. Because of resource sharing, popularization of multimedia equipment and convenience of mobile devices nowadays, people gradually break the existing framework about the definition of space. Space and residence are no longer limited to traditional family functional categories, but have more demands and choices. Similarly in this kind life circle, you and me can use more opportunities to explore possibilities of space and material and even the original demands of space material. On one hand, we can pursue extreme original and rough appearances of materials and track back to the deepest connotation of the space; on the other hand, we can surpass space categories and figure out individual thoughts. So far, people's propositions to space have been changed greatly. As a new display of life culture, liberating the space and hard framework bound of the space has become a "tacit" fact for many people.

With modern explosive culture and the clarity of personal consciousness, our orientations to space are gradually towards individualism and not limited to one form or style, and the functional utilizations of residential spaces become relatively diversified. In the process of thinking space, we no longer use a single style to divide individual exclusive space field. The display of personalized space can contain more informal style element collections and customized products assemblies. In other words, it is a kind of aesthetic display of personal life culture.

Facing with diverse, fast and complicated life culture, designers in thinking space in the future need to be more rational to face with these strong changes, lead out personalities of space protagonist by a thinking wisdom in main axis type, including leading imaginations and demands, integrate comprehensively by a kind of high unity and create a proper living space.

Details in projects by THEE design group are considered constantly in the design process and weighed and chosen constantly to get the final appearance. The design purpose is that we hope to mark a strong and vital postil for the target and space. For us, this postil must have a unique and distinctive image. What you see is what you get, it must satisfy the viewers in some degrees visually and spiritually. Original vitality is always a very important and meaningful thing for our clients and is also the value which we have been pursued all the time.

EXPLORING INITIAL PURSUIT, PLAYING ORIGINAL VALUE
探究原始诉求 发挥原创价值

身处 21 世纪的新生活里，我们以往习以为常的社会主义形式渐趋单薄化，进而取代的个人思维日趋成熟。现今的生活中因为资源共享、多媒体设备普及、行动装置的便捷等，人们对于空间的定义也逐渐打破原有的框架。空间与住所，不再只是局限于传统家庭的功能性范畴，而是有了更多的诉求和选择。同样在生活圈中的你我，反而会利用更多的机会去探讨空间与材质的可能性，甚至是对于空间材料的原始诉求，一方面追逐材质极致的原始粗犷样貌，回溯空间原本最深层的蕴含基底，一方面或超越空间范畴反推个人思维。演变至今，人们对于空间的命题已经发生了大幅的走向，解放空间与空间的硬式框架束缚，作为一种新的生活文化展现，已是越来越多人"心照不宣"的事实。

随着现代爆炸性文化充斥，加上个人意识的明朗化，我们对于空间的定位，也逐渐走向个人主义，不再受限于一种形式或风格，对于住宅空间的机能使用也变得相对多元化。在思考空间过程中，也不再只是单一使用某种风格来划分个人专属的空间场域，个人化空间的呈现能够包含更多非制式风格元素的汇集与订制品的集合，即个人生活文化的一种美学展现。

面对多元快速而复杂的生活文化，未来在思考空间的设计工作者，更需要沉着理性地面对这些强烈的变化，并且以一种主轴式的思考智慧，引导空间主角的性格内化导出。包括引导想象、引导诉求等，并以一种高度的整合力，全面性的汇整，进而创造出适切的生活空间。

在惹雅案例作品中所见到的各项细节，都是在设计过程中经过不断琢磨，不断衡量取舍，才让最终样貌得以呈现。设计的初衷在于，希望为对象与空间画下一个强烈而具生命力的批注。对我们而言，这个批注必须具有独特性及鲜明图像性，所见即所得，在视觉上与精神上都必须让观者得到一定程度的满足与满意。原创的生命力，这件事对我们与客户而言，都是极其重要而且极具意义的事情，也是我们一直以来诉求的价值。

THEE design group/ Chang Kai
惹雅设计 / 张凯

CONTENTS
目录

NATURAL STYLE · 自然风

008	风赋 WINDY SPACE	016	穿越·苏园 TRAVERSING SUZHOU GARDEN
026	灰阶上下 FLIGHT OF GRAY	034	原境·回廊 PURELY CORRIDOR
040	馥筑 FRAGRANCE	046	韵 x 律 MELODY CHARM
052	隽雅静好 ELEGANCE AND TRANQUILITY	058	遇见优雅的 Mr. 邦德 MEET JAMES BOND
066	艺墅 LIFE WITH ART	074	闲庭秀木 LEISURE YARD WITH BEAUTIFUL TREES
082	谱出流畅动线 改写空间新定义 COMPOSING FLUENT KINETONEMA, REDEFINING THE SPACE	090	贴心机能宅 THOUGHTFUL AND FUNCTIONAL RESIDENCE
096	透天美墅 TRANSPARENT BEAUTIFUL VILLA	102	风和·无相 GENTLE BREEZE WITHOUT FORM
106	简约生活居所 CONCISE LIVING RESIDENCE	114	坐拥度假景观宅 HOLIDAY LANDSCAPE RESIDENCE
120	传承世代的家族居所 展开疗愈的新生活 INHERITING FAMILY RESIDENCE, BEGINNING A SANATIVE NEW LIFE		

LIGHT LUXURIOUS STYLE · 轻奢风

134	缔造古典都会居所 CREATING CLASSICAL METROPOLIS RESIDENCE	142	时间的温度 TEMPERATURE OF TIME

164 木石重奏 金石为开 ENSEMBLE OF WOOD AND STONE BECAUSE OF A WILLING HEART	**174** 极致演绎家居艺术生活 EXTREMELY DEDUCING HOUSEHOLD ART LIFE
180 细致与冷抽象 DETAIL AND COOL ABSTRACTION	**186** 华丽城堡 GORGEOUS CASTLE
194 极致格调 THE EXTREME STYLE	**202** 帝之苑 THE IMPERIAL RESIDENCE
208 线语·框载 LINE EXPRESSION AND FRAME CARRIER	**214** 漫·艺 ROMANCE AND ART
220 享田园快意生活 ENJOYING RURAL AND LEISURE LIFE	**230** 感受自然脉搏的天井别墅 FEELING NATURAL PULSE IN THE COURTYARD VILLA
238 时尚聚落 FASHION SETTLEMENT	**244** 新东方 心奢华 NEW EAST AND HEART LUXURY
254 沐光 MU-RAY	**260** 奢华魅力 LUXURIOUS CHARM

INDUSTRIAL STYLE · 工业风

268 设计 x 想象：住进航海王的家 DESIGN & IMAGINATION, LIVING IN THE HOME OF ONE PIECE	**278** 漫舞空间谱奏双人圆舞曲 DANCING SPACE PLAYS A DOUBLE WALTZ
284 藏在禅风基调里的活泼因子 ACTIVE FACTORS HIDDEN IN THE ZEN-LIKE TONE	**294** 暖灰的美学情境 WARM GRAY AESTHETIC SITUATION
300 慢行 TIMELESS	**306** 重彩 VIVID COLOR
314 记忆，任意门 MEMORY, DIMENSION DOOR	

DESIGN CONCEPT　设计理念
SPACE PLANNING　空间规划
DECORATIVE MATERIALS　装饰材料
NATURAL LIGHTING　自然采光
NATURAL COLORS　自然色彩
INTERIOR VIRESCENCE　室内绿化
BRINGING SCENERY INTO HOUSE　引景入室

NATURAL STYLE
自然风

presenting plain and humanity, enjoying tranquility and comfort

谱写质朴人文，享受恬淡舒适

大道至简
SIMPLIFYING PROFOUND PHILOSOPHY

自然风 / NATURAL STYLE

WINDY SPACE
风赋

◎ DESIGN CONCEPT 设计理念

"Integrating environment into design, integrating design into art" is the design spirit of this project. The designers start with the concept of "wind" from environmental analysis, design image transformation in early stage to the combination of soft ware cloth and artistic home later. The main tones of the space are black, gray and wood color, creating comfortable and texture living space. At the same time red and yellow activate colors of part of the space.

"环境融入设计，设计融入艺术"是本案的设计精神，设计初期从环境分析、设计意象转化发展到后期设计的软件布艺及艺术家的融合都以"风"的概念为出发点。空间以黑色、灰色和原木色为主色调，营造出舒适富有质感的居住空间，同时介入红色和黄色，起到活跃局部空间的作用。

Design company ｜ Zline Concept Design
设计公司 ｜ 泽林空间设计

Designers ｜ KiLin, Laurie
设 计 师 ｜ 林星潍、董茗萱

Photographer ｜ Joey Liu
摄 影 师 ｜ 刘欣业

Location ｜ Kaohsiung, Taiwan
项目地点 ｜ 台湾高雄

Area ｜ 422m²
项目面积 ｜ 422m²

Main materials ｜ natural solid wood veneer, imitated wood grain quartz tile, NEOLITH Spanish stone brick, spray lacquer, etc.
主要材料 ｜ 天然实木皮、仿木纹石英地砖、NEOLITH西班牙耐丽石薄板砖、喷漆等

◎ DECORATIVE MATERIALS 装饰材料

In order to create magnificent and open living room in the first floor, the sofa background deliberately enlarges the scale of design elements and uses three-centimeter dark gray tile slices to present light and rhythmic wall modeling in irregular collage method. The two sides continue aluminum grille light and shadow of the architectural facade to present another superficial temperature by wood veneer modeling. The ceiling chooses natural grains and white walnut solid wood veneer to highlight the rhythm of the space. The second interlayer had a room with extra spaces. After negotiation, this room is designed into a whole open space as the family interaction space, which enhances the ventilation and day lighting of the living room.

一楼客厅为了营造空间的大气与开阔，在沙发背墙上刻意放大了设计元素的尺度，运用 3mm 厚深灰色的耐丽石磁砖薄片以不规则角度错落拼贴的手法呈现轻巧并具有漂浮律动感的壁面造型，两侧则延续建筑外观的铝隔栅光影以实木皮造型表现令一种表面温度，天花板特别选用自然纹路且对比分明的白胡桃实木皮来突显空间的律动感。二层夹层原有格局有一间房间，但房间外原有一畸零空间运用性不佳，经沟通后将此房间打开成为一整区开放空间，作为家庭互动空间，同时为挑高的客厅采光通风性能更加分。

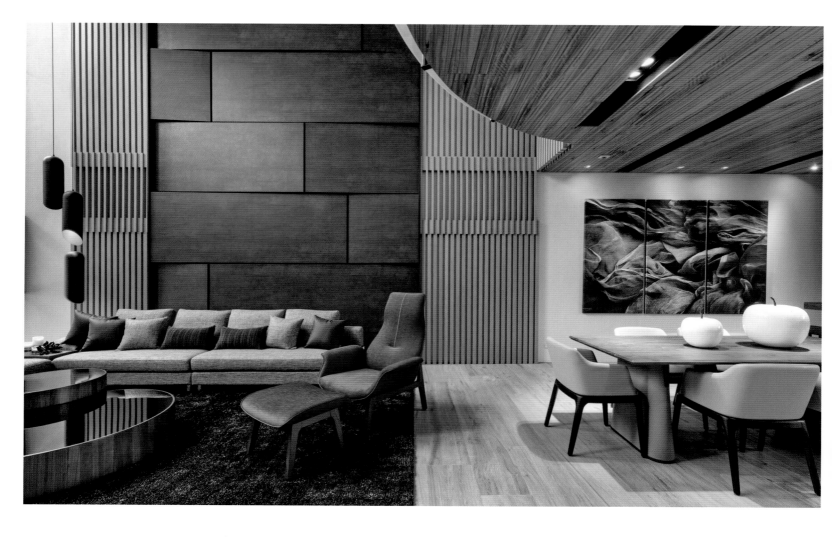

◎ NATURAL LIGHTING 自然采光

The double-deck living room uses large areas of glass window to bring outside lights and natural scenery into interior. The dining room also uses glass window to echo with the day lighting of living room. The living room, dining room and kitchen connect each other openly, which makes the interior more spacious and magnificent and let air and light flow freely to achieve free communication with the space. The family activity area in the second floor adopts open pattern to achieve perfect effects of day lighting and ventilation.

双层挑高客厅运用大面积玻璃窗将室外光线、风景自然引入室内，餐厅同时也使用玻璃窗，与客厅采光实现对面呼应。客厅、餐厅和厨房呈开放式相连，不仅使室内更显开阔大气，同时让空气和光线能自由流动，实现空间自由对话。二层家庭活动区间则以开放式的格局实现采光和通风的最佳效果。

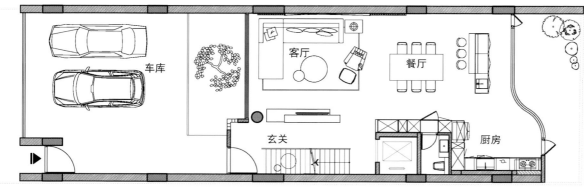

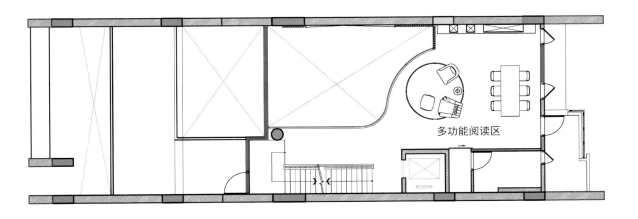

多功能阅读区

大道至简
SIMPLIFYING PROFOUND PHILOSOPHY

自然风 / NATURAL STYLE

TRAVERSING SUZHOU GARDEN
穿越·苏园

◎ DESIGN CONCEPT 设计理念

Based on ancient colors and modern charms of Suzhou Garden, the designer constantly thinks of the ancient Venetian waterway historic house. Overlapping stones, winding bridges, tranquil paths, antique woods, pavilions, corridor walls, beautiful windows, halls, lofts and gray tiles are as if in front of our eyes. The designer injects this kind of garden charm with interlaced senses of time into this modern space, composing a poetic song full of urban fashion sense.

设计理念脱胎于苏州园林的古色今韵，设计师对这个古色古香的威尼斯式水道古厝念念不忘，迭石、曲桥、幽境、古木、亭台、廊壁、花窗、厅堂、阁楼、青瓦仿佛仍在眼前浮现。设计师将这充满时代交错感的园林韵色注入这个现代空间，谱写出充满都市时尚感的诗歌赋曲。

Design company ｜ THEE design group
设计公司 ｜ 惹雅设计

Designer ｜ Chang Kai
设 计 师 ｜ 张凯

Photographer ｜ WEI
摄 影 师 ｜ WEI

Location ｜ Suzhou, Jiangsu
项目地点 ｜ 江苏苏州

Area ｜ 430m²
项目面积 ｜ 430m²

Main materials ｜ wood veneer, stone, black iron, imitated rock brick, etc.
主要材料 ｜ 木皮、石材、黑铁、仿岩砖等

◎ DECORATIVE MATERIALS 装饰材料

Neat modern style becomes the aesthetic performance of the space. Under the collocations of black iron, stone and wood grain, bright texture is obvious in the quiet atmosphere as if fancy landscapes with strange peaks and rocks setting off mountain and water in Suzhou Garden. The arrangement of large area of stone and wood creates a grand visual scale, which is magnificent with aesthetic style. Iron parts outline ladder modeling and walk in the space, exquisitely depicting transformation and change of the field in the space. The dining room and kitchen are separated by wood screen, slightly dividing functional fields. The central screen uses hollow structure, bringing natural lights from kitchen side door and spreading to the dining area along with the gap, which shows a clean bright vision.

利落的现代风格成为空间的美学展演，黑铁、石材与木纹肌理的搭配下，鲜明的纹理在一片寂静氛围里呼之欲出，彷佛苏州园林里一幕幕奇峰异石与山水相映的奇幻景致，大面积石材与木作的安排，则创造出壮阔的视觉尺度，隐然大气又不失美学风范；铁件勾勒出阶梯造型且游走在空间之中，细腻地描绘出该领域在空间里的转换与变化；餐厅与厨房之间则借由木作屏风，些微划分出领域机能，屏隔中央以镂空的结构呈现，从厨房侧门带来的自然光，也随着缝隙蔓延至用餐领域，展现洁净明亮的视野。

◎ SPACE PLANNING 空间规划

The designer turns Suzhou Garden scenery into visible indoor things. Every step forward, whether magnificent living room, simple and neat kitchen or warm and elegant bedroom, all can make you feel antique charms of Suzhou Garden. The space is well-organized with practical functions and aesthetics.

设计师将苏州园林美景转化为室内可观的一物一具，每前进一步，无论是气势壮阔的客厅，还是简单利落的厨房，抑或是温馨素雅的卧室，都感受到苏州园林的古色古香气韵。空间设置齐全，兼具实用与美感。

◎ NATURAL COLORS 自然色彩

The pleasant gray sofa, simple and plain wood color tea table and ink black shelf in the living room intersect cold and warmth, creating a rich visual effect. The bedroom uses soft white curtains to bring clean and bright temperament for the space, endowing it with carefree and comfortable atmosphere.

客厅宜人的灰色沙发、简朴不造作的木色茶几、冷硬浓重的墨黑书架，冷暖交汇，创造出丰富饱满的视觉效果。卧室采用轻柔的白色纱窗，为空间带来白净明亮的气质，赋予空间悠然舒畅的氛围。

大道至简
SIMPLIFYING PROFOUND PHILOSOPHY

自然风 / NATURAL STYLE

FLIGHT OF GRAY
灰阶上下

◎ DESIGN CONCEPT 设计理念

There is an open mouth of the wood cabinet in the space, and a pedal comes out from it. Metal suspensions are used throughout upstairs and downstairs, along with the leading of stainless steel light band upstairs and downstairs, which not only transitions and connects two different lives upstairs and downstairs, but also interprets different lifestyles and personalities of residents. Up and down in the same gray tone present two different yet necessary orientations of life and endow life with different satisfactions.

空间中的木柜被开了个口，有踏板从柜内长出，以金属悬吊贯穿楼上楼下，沿着不锈钢光带的指引上楼下楼，过渡的、连结的不仅是楼上楼下两个不同的生活空间，也是居住者对不同生活方式及个性的诠释。上与下，在同样的灰色基调下，呈现出生活的两种不同却必需的面向，也让生活获得了不同的满足。

Design company ｜ Create+Think Design Studio
设计公司 ｜ 创研空间

Designer ｜ Arthur Ho
设 计 师 ｜ 何俊宏

Location ｜ Taipei, Taiwan
项目地点 ｜ 台湾台北

Photographer ｜ Kuomin Lee
摄 影 师 ｜ 李国民

Area ｜ 160m²
项目面积 ｜ 160m²

Main materials ｜ sound-absorbing cotton, walnut wood floor, wood veneer, iron part, etc.
主要材料 ｜ 吸音棉、胡桃木地板、木皮、铁件等

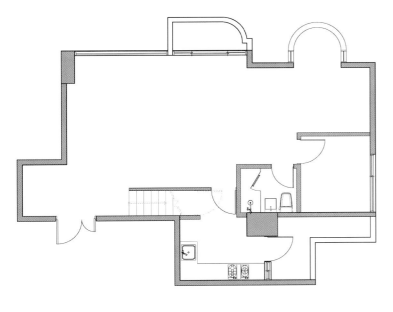

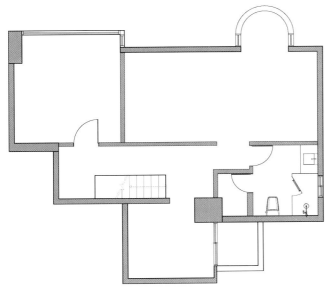

◎ DECORATIVE MATERIALS 装饰材料

The space is presented by gray tone in different ways and produces different textures according to different material needs. In order to pursue perfect reflection of sound, public area uses different concave and convex three-dimensional texture materials. In order to pursue cleanness, the dining room and kitchen adopt professional kitchen stainless steel as main material, which is convenient to clean and brings high quality texture of modern crafts.

空间以不同方式的灰呈现，依据不同需求材质产生不同肌理。聆听空间的公共区域为求声音最佳的反射，采用不同凹凸立体质感的面材；餐厨区域为求洁净，以专业厨房用的不锈钢为主要材质，方面清洗的同时，带来现代工艺的高级质感。

◎ NATURAL COLORS 自然色彩

The downstairs appreciation space is paved with large area of gray and wood color. The elegant gray and warm wood perform a beautiful encounter in the space. Upstairs uses the same gray, but it uses another hand daubed cement gray which can easily make people deposit, which is very suitable for creating comfortable and tranquil atmosphere for the rest space.

楼下的鉴赏空间以大片灰色和木色铺就，淡雅的灰与温暖踏实的木色在空间中上演一场美丽的邂逅。楼上空间同样以灰色呈现，但是另一种手作涂抹的、让人易于沉淀的水泥灰，非常适合营造休憩空间里舒适静谧的氛围。

◎ SPACE PLANNING 空间规划

According to functions, the space is divided into two areas upstairs and downstairs, which are connected by a suspended ladder. Upstairs can precipitate body and mind, where residents can rest, bath and read; downstairs is an appreciation space, where residents can listen to music to receive auditory stimulation and taste food and wine to relax. Even on the same floor, there is a contrast of two kinds of atmosphere.

空间按机能分为楼上楼下两个区域,两个空间由悬吊式阶梯连接。楼上是可以沉淀身心的所在,居住者可在这里休息、沐浴、阅读;楼下是个鉴赏空间,既可以在这聆听音乐,接受听觉的刺激;也可品尝美食好酒,享受美食好酒带来的放松愉悦。虽然在同一楼层,却形成反差极大的两种氛围。

PURELY CORRIDOR
原境·回廊

◎ DESIGN CONCEPT 设计理念

This is a "congenitally deficient" space with a long corridor in the middle of the living room and bedroom. The congenital conditions of the space make the corridor become the essential element to link public area with private space, trying to endow the certainly existed corridor with stronger place spirit. The corridor spreads, extends and becomes an indispensable medium and main design axis of the entire space, which connects the space in a unified tone.

The sunshine is quiet and comfortable and the time is tranquil and peaceful. Sun lights from the shutters sparkle in the ground and warm body and heart. This is the surprise from the designer to owner. Home is not just a space but an existence with temperature.

这是一个"先天不足"的空间，一条长长的廊道隔在客厅与卧室中间。由于此空间的先天条件，使得廊道成为链接公共区域及私密性空间之必要元素，试图让必然存在的廊道空间赋予其更加强烈的场所精神，让此一廊道扩散、延伸，并成为整体空间不可或缺的介质及设计主轴，串联空间，统一调性。

阳光安适，岁月静好。透过层层百叶窗传递进来的日光，层层叠叠，打在地上，暖在身心。这便是设计师希望传递给屋主的惊喜！家，不仅仅是一个场所，它是有温度的存在。

Project name ｜ Impractical Illusion
项目名称｜非非想

Design company ｜ DESIGN APTMENT.
设计公司｜近境制作

Designer ｜ TT
设 计 师｜唐忠汉

Photographer ｜ MW PHOTO INC
摄 影 师｜岑修贤摄影工作室

Location ｜ Taipei, Taiwan
项目地点｜台湾台北

Area ｜ 225m²
项目面积｜225m²

Main materials ｜ wood veneer, PANDOMO, iron part, tile, spray lacquer, fabric, stone, etc.
主要材料｜木皮、盘多魔、铁件、磁砖、喷漆、绷布、石材等

◎ NATURAL COLORS 自然色彩

Excluding foppery colors and returning to initial hues are the purposes of the designer. There is a kind of taste without decoration that is simple and pure, yet has a warming power. When each material is allowed to return to its original intention, is presented appropriately and can coordinate with each other, the natural and unpretentious atmosphere of space can be expressed. Light and shadow is not only the superficies of shining on objects, but also the essence of illustrating spatial heritage.

设计师以摈除矫饰的色彩，回归到最初的色调为宗旨。空间架构，不加修饰，质朴、纯粹，却有一股温暖的力量。使材质归乎于初衷，当各个媒材得到适当的发挥，能为其所用，能相辅相成，空间自然而然以不造作的氛围呈现流露。光影，不仅仅是照射物体的表态，却也是刻画空间底蕴的本质。

◎ SPACE PLANNING 空间规划

There is a kind of space that intertwines time, memory and longing to form beautiful and touching thoughts. A central axis penetrates through the public and private domains to separate them, yet utilizes water finish veneers and wooden floorings to connect the vital interactions of the overall space. This is the connection between domains and the connection of sensations.

有一种空间，交错时光，回味、想念，一道美好又感动的思念。由中轴线贯穿公领域及私领域，相互分割渗透，以水染木皮与木质地板串联出整体空间的重要联系。这是空间与空间的串连，亦是感情上的连结。

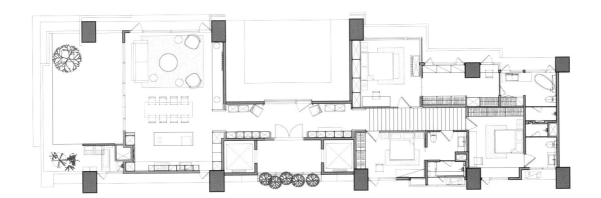

◎ DECORATIVE MATERIALS 装饰材料

There is a kind of occasion where you can disarm yourself and grow with a series of soliloquy and self-examination. Taste the quietness constructed by the original forms of the materials and enjoy the mild simplicity created by the surroundings of space. After precipitation, face the beginning of the dialogue between oneself and the space calmly. Different from the linking to the feeling of the above-mentioned corridor, this corridor is the emotional memory that symbolizes pureness. Getting rid of forms, the plain and straight way presents an experience different from the previous style.

　　有一种场合，卸下武装，自我、对话，一场内省释放的成长。仔细品味材质原始姿态下所建构而成的安谧，品茗空间围绕形塑出的温润朴质。沉淀后，从容的面对空间与自身之相互对话的滥觞。有别于空间脱离的连结，却创造出回归单纯的情绪记忆。跳脱形式表现，以朴实直述的方式，呈现不同于以往的风格体验。

大道至简
SIMPLIFYING PROFOUND PHILOSOPHY

自然风 / NATURAL STYLE

FRAGRANCE
馥筑

◎ DESIGN CONCEPT 设计理念

This project is located in the tall building on Po-ai Road in Kaohsiung. The owner loves views here and hope to see the scenery and wash body and mind when going out early and coming back late. So the design of public areas becomes the focus of the whole space. The designer redistributes space scale, adjusts the most comfortable planning of visual sense and sets window scene in a long form with flowing images to feel the dialogue between light and shadow and space. When the outdoor sunshine passing through the shutters, it is like flowers falling in the space. The ground is silence, which adds an elegant flavor to the clean and plain modern space. The designer uses modern materials to present restrained manner of the whole space, creating an extraordinary reception manner.

此案位于高雄博爱路上的高楼层住宅，业主喜爱这里的居住视野，希望早起或晚归时，能尽览景色，洗涤身心。于是公共区域的设计成为整个空间的重点，重新分配空间上的比例，调整出视觉感最舒适的规划，让窗景成为长型的量体，变成流动的画面，就能感受到光影与空间的对话。当户外暖阳从百叶窗穿过，如簌簌落花铺满室内，满地寂静，替干净、无华的现代空间增添一丝优雅光意。设计师用具有现代气息的材质用料铺陈出空间整体的内敛气度，营造非凡的迎宾气度。

Design company ｜ CJ INTERIOR
设计公司 ｜ 长景国际设计有限公司

Designer ｜ Chris Wu
设 计 师 ｜ 吴冠谚

Photographer ｜ AR-HER KUO
摄 影 师 ｜ AR-HER KUO

Location ｜ Kaohsiung, Taiwan
项目地点 ｜ 台湾高雄

Area ｜ 261m²
项目面积 ｜ 261m²

Main materials ｜ stone, iron part, glass, titanium, stainless steel, solid wood veneer, etc.
主要材料 ｜ 石材、铁件、玻璃、镀钛、不锈钢、实木皮等

◎ SPACE PLANNING 空间规划

The design starts from porch, using line extension of grille into living room to lead the owner to enjoy custom-made space configuration. The right is rectangular TV collecting cabinet and the left is rectangular display cabinet, which divides public areas and enlarges the space. Design vocabularies use long measure body and vertical lines to shape space expressions and extend a new order. The hanging cabinet uses iron part and leather, with marbles in the floor. Stones connect the space and extend to the dining room, creating a low-key and restrained texture. The whole space wood floor and shutters warm the space, depositing a humanistic texture.

设计的起点由玄关开始，利用隔栅的线条延伸进客厅，带领业主去领略量身定作的空间配置，右侧为长型的电视收纳柜，左侧也是长型的展示柜，让靠边的柜体划分出公领域的区域，使空间感极大化。设计语汇采用了长型的量体与垂直的线条交错来形塑空间的表情，延伸出新的秩序；悬吊的柜体材质上则用铁件及皮革表现，搭配大理石当底，让石材串连空间延伸至餐厅区域，营造低调的内敛质感，加上全室木地板及木百叶温润空间，沉淀出人文质感。

◎ DECORATIVE MATERIALS 装饰材料

This project uses stone, precise grain stainless steel and glass to create rich sense of layering. Warm wood presents warmth in concise and neat lines. The bathroom background wall is covered with smoky gray marbles which reveals space manner and interprets touching texture when enhancing with the white bath.

此案运用了石材、精密纹理的不锈钢及玻璃等材质，创造丰富层次感，温润的木质则在简洁利落的线条中，注入暖度。浴室以烟灰色大理石打造的墙壁背景，石材肌理彰显居室气度，与白色浴缸相互辉映，诠释出动人质感。

◎ NATURAL COLORS 自然色彩

The living room sets elegant gray and sedate black as the tone. The orange seat creates waves, which is like small stone when thrown in the water. The master bedroom uses gray tone, fabric cloth and arc modeling grille to soften the overall vision and plan double dressing rooms, which conforms to the new couple's expectations and becomes their wanted space.

客厅以优雅的灰色和稳重的黑为基调，点缀的橙色座椅如投入水中的小石子引起波澜。主卧室则以灰阶色调搭配，同时采用布质裱布、圆弧的造型格栅，柔化了整体视觉，并规划出双更衣室，符合新婚夫妻的期望，成为心之所向的居所。

大道至简
SIMPLIFYING PROFOUND PHILOSOPHY

自然风 / NATURAL STYLE

MELODY CHARM
韵 x 律

◎ DESIGN CONCEPT 设计理念

Entering the door, mellow, melodious and warm chord begins to play. Following the rhythmic melody of arcs, footsteps can be light. Entering the hall, the music turns to gentle and stretching adagio, slowly flowing flavors of the space and finding an eternal and harmonious order in the vertical and horizontal three-dimensional space. In this space whether orderly geometry or flowing organic line, it is like the rhythm and flow of music and brings life energy to the space just like light and shadow. The gentle and stretching adagio and fluent and melodious allegro endow the space with different flavors. In addition with artistic performance of the rhythm, the space can be broader and deeper.

一进家门，圆润悠扬暖白的和弦旋即奏起，跟随弧线律动的旋律，脚步也随着轻盈起来。进入门厅后音乐开始转化成和缓舒展的慢板，缓缓流动出空间的韵味，并在三度空间的垂直与水平向度中，找到永恒和谐的秩序。在这个空间里不论是有秩序的几何或是流动的有机线条，都像是音乐的韵律性与流动感，和光影一样，为空间带来生命的动能，时而和缓舒展的慢板，时而流畅悠扬的快板，赋予空间不同的韵味，加上韵律的艺术性表现，可让空间表现更加宽广与深邃。

Design company ｜ DINGRUI Design Studio
设计公司 ｜ 鼎睿设计有限公司

Designer ｜ Dingrui Tai
设 计 师 ｜ 戴鼎睿

Photographer ｜ Figure x Lee Kuo-Min Studio
摄 影 师 ｜ 图起乘李国民影像事务所

Location ｜ Taipei, Taiwan
项目地点 ｜ 台湾台北

Area ｜ 325m²
项目面积 ｜ 325m²

Main materials ｜ wood veneer, iron part, cedar, paint glass, marble, Zinn Bernie stone, oak wood floor, etc.
主要材料 ｜ 木皮、铁件、香杉、烤漆玻璃、大理石材、辛伯尼石材、橡木地板等

◎ INTERIOR VIRESCENCE 室内绿化

People's yearning to nature has never decreased. Being in the bustle city and closed living space, only adding green plants can comfort the heart which obsesses nature while never gets it. Green plants in the entrance porch and small bonsais on the tea table can make you feel greenery and vitality after entering the door. There are also green plants in the living room, dining room and study. There is a bottle of flowers on the bedside table of the bedroom, which emits light fragrance randomly. Even in the bathroom, there are all kinds of distinct small bonsais, which is green and fragrant. Living here for a long time, you must be happy and pleasant.

人们对大自然的向往之情从来都未减少，身处都市的繁华与封闭的居住空间，唯有在室内加入绿色植物，以抚慰那迷恋自然而不得的心。入门玄关的植物盆栽和茶几面的小盆景，入门便能感受绿意生机。客厅、餐厅、书房也随处可见绿植，卧室则在床头柜上轻放一瓶插花，简单随意中透出淡淡芬香，就连浴室中也可见到各种特色的小盆栽，绿意芬芳，环绕室内。久居其间，心悦神怡。

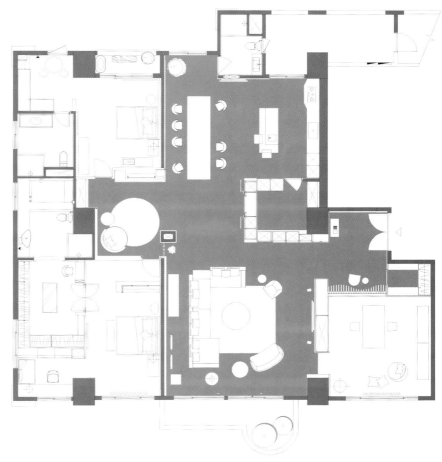

◎ NATURAL COLORS 自然色彩

The slowly flowing rhythm of the space produces a comfortable and harmonious living beauty. The main space sets gray as the tone with white ceiling, which manifests balanced beauty of color in the dark and light collocation. White cleanness and elegance, gray low key and texture and orange and red sofa and chair add visual focuses. Green plants in space ripple out energetic vitalities. Even being indoor, there are always green plants.

空间缓缓流动的韵律激荡出舒适和谐的居住美感，以灰色为基调的主空间，镶嵌着白色天花，深浅搭配中流露出均衡的色彩之美。白色的明净优雅，灰色的低调质感，同时用橙色、红色的沙发或单椅，为之增添视觉上的聚焦点。空间中的绿色植物，荡漾出昂扬的生机，即使身处室内，也常有绿植相伴。

大道至简
SIMPLIFYING PROFOUND PHILOSOPHY

自然风 / NATURAL STYLE

ELEGANCE AND TRANQUILITY
隽雅静好

◎ DESIGN CONCEPT 设计理念

Absorbing the essences of modern style and Oriental elegance, the designer uses rugged stones in this concise and restrained 150 square meters residence to promote its artistic level with grilles. Vertical orders of lights and shadows, lateral ink trails and plain scenery intersect harmoniously. The appropriate white space wall reaches an appropriate contrast and a perfect balance in the heavy gorgeous neutral color.

汲取现代风格与东方典雅的精粹，设计师将嶙峋的岩墨石质，收融于简洁内敛的150平方米居宅，进而提升居宅的艺术层次，并借由格栅疏漏。梳络光影的垂直秩序，与横向的墨染笔势、一脉平原景色和谐交织；而适度的留白墙面，更于浓郁绮丽的中性色彩之中，觅得恰如其分的对比与完美平衡。

Project name ｜ Humanities
项目名称｜臻邸 敛隐台中

Design company ｜ Snuper Design
设计公司｜大雄设计

Designer ｜ Cheng Wei Lin
设 计 师｜林政纬

Photographer ｜ Lee Kuo-Min & Snuper
摄 影 师｜李国民 & Snuper

Location ｜ Taipei, Taiwan
项目地点｜台湾台北

Area ｜ 150m²
项目面积｜150m²

Main materials ｜ stone, super durable floor, wood veneer, iron part, etc.
主要材料｜石材、超耐磨地板、木皮、铁件等

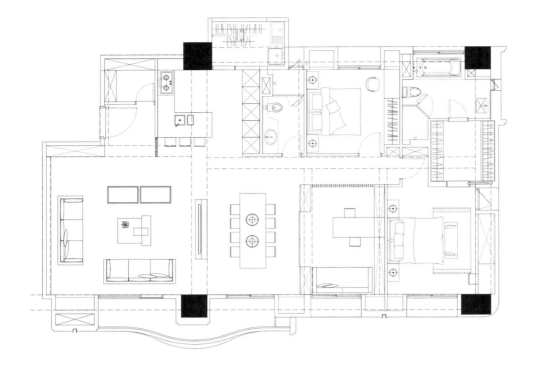

◎ DECORATIVE MATERIALS
装饰材料

Entering from the porch, the screen partition combines profound and calm wood, stone color and grille lines to create a rich and restrained atmosphere, collocating with the stone texture like ink strokes and winding cabinet door pieces, which quietly presents extraordinary and luxurious tastes in the picture arrangement of textures and details.

自玄关处步入，屏隔揉合深邃沉稳的木、石色泽与格栅线条，营造醇厚内敛的氛围印象，搭配如水墨笔触般的石质脉理，与一旁具蜿蜒曲度的柜体门片，在材质细节的摹画安排上，低调展露不凡的奢华品味。

◎ SPACE PLANNING
空间规划

The open space design makes the kinetonema more fluent and successfully integrates the living room, dining room and kitchen into a whole. There is no TV in the space. The poetic flow goes from B&O sound where songs come from and extends to the open dining table and kitchen. Concise lifestyle and high quality life present the communication with the environment, thinking of life essence one step back and open mind. The family story begins from the residents.

开放式的空间设计，是为了让空间动线更为流畅，成功地将客厅、餐厅、厨房等空间合而为一。空间中没有电视，诗性的流动沿窗走到了B&O音响，经典歌曲从这里发散，延伸到了开放餐桌与厨房。简约的生活方式，高质感的居住质量，展现的是与环境对话、退一步省思生活本质、开阔的人生胸怀，家的故事从居住者开始写起。

◎ NATURAL COLORS 自然色彩

With the kinetonema turning to the living room, it continues the deep color in the entrance. The natural texture of stones forms a magnificent visual side view. The open design connects kitchen and dining room, which creates a broad and free life vision. Through large areas of French widows near the sofa and dining table, you can see the distant wetland and plain, which skillfully embeds a fresh green scenery into the sedate manner. The master bedroom uses the same sedate wood tone and keeps capacious and comfortable space scale. The middle is dressing area, which forms a symmetric and harmonious vision with both sides of window views.

随着动线转折进入客厅领域，延续入口处的深邃色调，岩墨的自然脉理在此摊展为一幅磅礴大气的视觉端景，并以开放式设计连贯厨房及餐厅处，形构辽阔自在的生活视野。而透过沙发及餐桌旁的大面落地窗，则可远眺湿地平原，巧妙地在沉稳气度中镶嵌一方鲜活绿景。主卧室同样采用沉稳的木质色调，并保留宽敞舒适的空间尺度，中段安排为梳妆区，与两侧窗景形成对称的和谐视觉。

MEET JAMES BOND
遇见优雅的 Mr. 邦德

◎ DESIGN CONCEPT 设计理念

The designer is good at using advanced materials and devices which may look common and simple, grains and characteristics relying on material essences, orderly and careful arrangements and exquisite craft modeling , in addition with skilled color application techniques and appropriate lighting, to present extraordinary visual enjoyments and concisely express spiritual pursuits of the space. A brilliant place leaves a light aftertaste without lingering too long, then guiding the viewer to move forward.

设计师擅长将看上去平淡简单实则高级的材质和器具，或是依托材质原本的纹路及特性，或是经过有序精心地安排，或是讲究的工艺造型，加上娴熟的色彩运用技巧，恰当好处的灯光映衬烘托，呈现精彩非凡的视觉享受，言简意赅地表达空间的精神诉求。偶尔妙笔生花的一处，并不多做停留，留下淡淡余韵，即把观者目光引向更深处。

Design company ｜ THEE design group
设计公司 ｜ 惹雅设计

Designer ｜ Chang Kai
设 计 师 ｜ 张凯

Photographer ｜ WEI
摄 影 师 ｜ WEI

Location ｜ Taipei, Taiwan
项目地点 ｜ 台湾台北

Area ｜ 320m²
项目面积 ｜ 320m²

Main materials ｜ leather, iron part, titanium, black mirror, etc.
主要材料 ｜ 皮革、铁件、钛金属、黑镜等

◎ DECORATIVE MATERIALS 装饰材料

The project is like a legendary craft of Montblanc. The designer uses yuppie to make creations. Saddle process is as a soul prelude of the field. The orderly lines are flexible and steady steps in the space. Under the collision between leather and bold rough stone, the conflict visual appearance jumps ups and downs, agitating a conflict and harmonious new form.

　　项目如同 Montblanc 的传奇工艺，设计师以雅痞之名来挥霍创意，马鞍工艺作为场域的灵魂序曲，充满秩序的线条，是空间中灵活而沉稳的舞步表情；皮革与粗犷原石的相互碰撞之下，极具冲突性的视觉样貌，不安分地此起彼落地跳跃着，激荡出一种新型态的冲突和谐样貌。

◎ SPACE PLANNING 空间规划

Smooth marbles spread from the entrance porch to deeper interior, pulling out a style corridor, defining activity areas, such as living room, kitchen and dining room, and dividing rest areas of bedroom and bathroom, which links different functions and connects every space to produce interactive and flowing space effects.

光洁大理石从入口处的玄关开始蔓延，延伸至室内深处，拉出一条风格廊道，界定了如客厅、厨房、餐厅等活动领域，也将作为休息区的卧室、浴室等分割开来。既衔接起不同的领域机能，也串联起各个空间，产生互动且流动的空间效果。

◎ NATURAL COLORS 自然色彩

Colors of the home rely not on quantity but on whether they are collocated properly. In this project, the designer adopts pure and low brightness tones, coupled with texture furniture and craft. "Entering into the space with the wind, nourish things silently", the space texture is like delicate spring rain which quietly infects people.

家中的色彩不在于多，而是在于是否搭配得恰当好处。本案中，多采用纯净、低亮度的色调进行搭配，加上极具质感的家具和工艺，"随风潜入夜，润物细无声"，空间质感如同细腻的春雨悄无声息地将人感染。

大道至简
SIMPLIFYING PROFOUND PHILOSOPHY

自然风 / NATURAL STYLE

LIFE WITH ART
艺墅

◎ DESIGN CONCEPT 设计理念

Considering "tasting life", this project integrates sufficient natural lighting and natural building materials in the base to turn sunshine, air and water into unique decorations in the space. The water curtain wall piled by stones makes clanging lingering sound of water collisions around the space, which manifests natural scenery and magnificent touching. We advocate touching of design and resonance of aesthetics. Design is more art than life. Though designs, residents can create their own life tastes and memories in a long time with the space and can repeatedly taste artistic life in it.

本案设计以"品味生活"为考虑，让基地中充足的自然采光及天然建材相互结合，阳光、空气、水转译成空间中绝伦独有的装饰品。结合入口墙体以石皮堆砌而成的水幕墙，让水流潺潺碰撞后的铿锵余音缭绕于空间，让此适切地流露自然光景及大器的感动。我们主张设计的感动和美感的共鸣。设计不仅是生活更是艺术，透过设计，居住者在漫长相处空间上能创造自己的生活品味及回忆，亦能再三回味其中交织出的艺术生活。

Design company ｜ Zline Concept Design
设计公司 ｜ 泽林空间设计

Designers ｜ KiLin, Laurie
设 计 师 ｜ 林星潍、董茗萱

Photographer ｜ Joey Liu
摄 影 师 ｜ 刘欣业

Location ｜ Kaohsiung, Taiwan
项目地点 ｜ 台湾高雄

Area ｜ 789m²
项目面积 ｜ 789m²

Main materials ｜ natural solid wood veneer, stone, water mould paint, iron part, spray lacquer, etc.
主要材料 ｜ 天然实木皮、石材、清水模涂料、铁件、喷漆等

◎ NATURAL LIGHTING 自然采光

Entering the reception room in the first floor, three sides are embedded with glass windows. Sun lights naturally sparkle in the interior through the windows, which is warm and vernal, tranquil and beautiful. The living room and dining room in the second floor are set with the same glass windows to achieve day lighting. At the same time there is a glass partition between them, which increases flowing and penetration of lights. In the open and unified layout, white floors, walls and ceiling achieve perfect effect of natural day lighting.

进入一层会客厅，三面以玻璃镶嵌为窗，阳光透过窗户自然洒落在室内，温暖和煦，静谧美好。二层客厅和餐厅同样装置玻璃落地窗，实现室内自然采光效果，同时区域之间以玻璃门隔断，加强了光线的流动和穿透。室内开放式、一体化的布局，白色为主的地砖和墙面、天花，实现空间自然采光的最佳效果。

◎ DECORATIVE MATERIALS AND NATURAL COLORS 装饰材料、自然色彩

Dark natural steel brush wood veneer chair leans on the sofa background, collocating with vertical slope and indirect lighting, which presents light transmission and magnificence and calmness. TV wall matches black wood grain stone with leather boards, using color lumps to connect vertical kinetonema. The dining room matches yellow dining chairs with dark color table, forming a contrast of cold and warm. The custom-made red wine cabinet is embedded in the brown displaying space, which outlines the ideal home full of vitality.

　　深色的天然刚刷木皮椅靠沙发背墙，垂直配以立体斜面及间接灯光效果，呈现光影传递及大气沉稳，电视主墙运用黑木纹石搭配梯侧皮革裱板，以色块量感串连垂直动线，餐厅以黄色餐桌椅搭配深色餐桌，冷暖均衡对比，同时结合专业量身打造红酒柜，嵌入棕色系展示空间，勾勒出充满活力的理想家园。

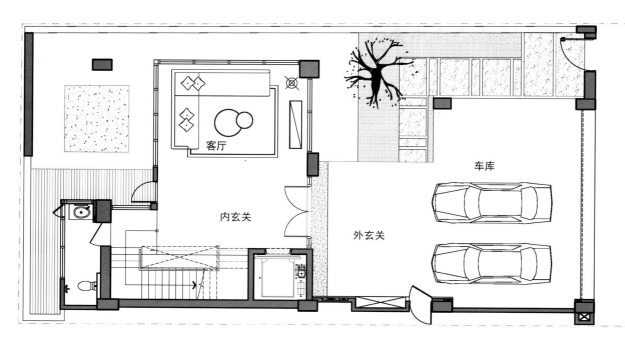

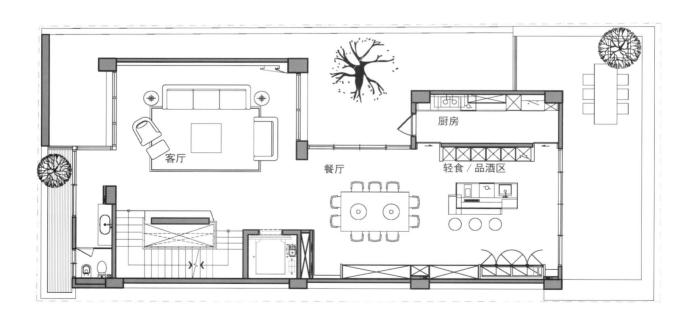

LEISURE YARD WITH BEAUTIFUL TREES
闲庭秀木

◎ DESIGN CONCEPT 设计理念

The porch entrance makes people feel the mansion style. Black cabinet and metal decorations use several strokes to outline luxurious manner of a mansion. Entering the living room, you can see details hidden in the basic tone. A lot of wood grains bring an extensive vision, collocating with black marble TV wall and well-arranged sofa and tea table, which presents an extraordinary manner of the living room. Large area of French window brings plentiful natural lights into interior, after white curtain "filter", creating clean and bright visual effects.

从玄关入门处起，就让人感受到豪宅风范，黑色小柜、金属装饰物，虽然只是寥寥几笔，也勾勒出属于大宅的奢华气度。步入客厅后，可见隐藏于基础色调的细节层次，大量木纹肌理带来延伸的视感，与黑色大理石电视背景墙，经过精心摆放的沙发、茶几一起，描绘出客厅的非凡气度。大型落地窗将充沛的自然光导入室内，经过白色窗纱的"过滤"，带来洁净明亮的视觉效果。

Project name ｜ Taoyuan Zhongyue Jiang Residence
项目名称｜桃园中悦江宅

Design company ｜ CHI-YI INTERIOR DESIGN STUDIO
设计公司｜奇逸空间设计

Designer ｜ Boshen Guo
设 计 师｜郭柏伸

Photographer ｜ SAM
摄 影 师｜SAM

Location ｜ Taoyuan, Taiwan
项目地点｜台湾桃园

Area ｜455m²
项目面积｜455m²

Main materials ｜ wood veneer, stone, iron part, glass, etc.
主要材料｜木皮、石材、铁件、玻璃等

◎ DECORATIVE MATERIALS 装饰材料

Warm wood wanders on the ceiling, lingers in the corridor and walks in the study, which presents a profound and sedate atmosphere. Glass with modern flavors brings sense of line and brightens tone of the space. Smooth marble floor presents mansion style and promotes space texture. Different materials bring different visual experience. Several kinds of materials collocate harmoniously and show their aesthetics maximally.

温润的木质游走在天花上，在走道上停留，在书房里漫步，铺陈出一室深沉稳重。具有现代气息的玻璃带来线条感的同时，点亮空间色调。光洁大理石地面则展现豪宅风范，提升空间质感。不同的材质带来不同的视觉体验，几种材质奇异地搭配协调，最大限度地发挥各自美感。

◎ SPACE PLANNING 空间规划

Looking from corridor to living room, you can see dignified and magnificent space layout. The designer uses grains of wood ceiling and black lamp furnishings to bring out layering of the plane and smoothly connect every space. Whether looking from living room to dining room or looking from dining room to living room, it can make residents feel interactive and flowing atmosphere.

由廊道望向客厅，可见端庄大气的空间轮廓，设计师借由木质天花的纹路及黑色灯饰带出平面的层次感，也流畅地串联起各空间领域。无论是由客厅望向餐厅，或是由餐厅望向客厅，都能让居者感受到互动而流动的氛围。

◎ NATURAL COLORS 自然色彩

The living room set calm wood color and profound black as the main tone, with the contrast and foil of different materials, adding rich layering into the space. The bedrooms continue the same black and white tone as the whole space. White ceiling, dark color background wall and floor and embedded cabinet lines create concise and texture rest spaces.

客厅以稳重的木色及深沉的黑色为空间主色调，加上不同材质的烘托对比，空间中无形中增加丰富的层次性。卧室延续整体空间的黑白色系，纯白的天花，利用深色系背景墙面、地面，及内嵌式橱柜线条，塑造简约而充满质感的休憩空间。

大道至简 / SIMPLIFYING PROFOUND PHILOSOPHY

自然风 / NATURAL STYLE

COMPOSING FLUENT KINETONEMA, REDEFINING THE SPACE
谱出流畅动线 改写空间新定义

◎ DESIGN CONCEPT 设计理念

The young couple who loves to try new things find Interplay Space Design Studio which has extreme creative energies, expects to use design methods jumping out of tradition, redefines space planning of this project through creative layout and reaches the maximum of function and beauty.

Starting from the porch, the ceiling arc as if a winding river, connects living room and dining room, fades into the corridor and leads the vision. Interplay Space Design Studio breaks the original rigid layout, uses arc lines to outline the ceiling of the porch, replaces one side real wall with glass and geometric lines to create vague sense of penetration, which makes the space get rid of closed feeling and endows life with complete and harmonious meaning. What's more, it removes the wall of the entrance corridor into a arc cabinet, creates a large scale opening by turn of angle, which eliminates sharp sense of the corner and broadens the scale of the corridor. The end of the corridor is equipped with large capacity of storage function, which meets collecting needs of the family.

喜爱尝试新事物的年轻屋主夫妇，找上极具创意能量的演拓空间设计，期望跳脱传统的设计手法，透过创新式格局规划，重新定义本案的空间脉络，达到机能与美感的最大值。

从玄关起始，犹如蜿蜒河流的天花弧线，贯穿客、餐厅后没入廊道，导引着视觉的行进。演拓空间设计突破原始格局的僵化，利用圆弧线条勾勒玄关天花面，并挖空一侧实墙，以玻璃与几何线条取代，创造出隐约的穿透感，让空间摆脱封闭，赋予生活的圆融意义。此外，拆除入室廊道墙面，规划为圆弧形柜体，并借由角度转折创造大尺度开口，消弭转角锐利感之余，亦拓宽廊道尺度，而廊道尾端则备有大容量的储藏室机能，满足屋主一家的收纳所需。

Project name ｜ Confluence
项目名称 ｜ 汇流

Design company ｜ Interplay Space Design Studio
设计公司 ｜ 演拓空间设计

Designers ｜ Ted, Ycy
设 计 师 ｜ 张德良、殷崇渊

Photographer ｜ KyleYu Photo Studio / Yuwei Huang
摄 影 师 ｜ 游宏祥摄影工作室/黄钰威

Location ｜ Hsinchu, Taiwan
项目地点 ｜ 台湾新竹

Area ｜ 210m^2
项目面积 ｜ 210m^2

Main materials ｜ black gold peak marble, silver white dragon marble, silver fox marble, wood veneer, wood veneer dying treatment, laminate hard plastic sheet, iron part spray lacquer, gray glass, dark glass, paint glass, imitated mould tile, mosaic tile, etc.
主要材料 ｜ 黑金锋大理石、银白龙大理石、银狐大理石、木皮、木皮染色处理、美耐板、铁件喷漆、灰玻璃、墨镜、烤漆玻璃、仿模板磁砖、马赛克磁砖等

◎ DECORATIVE MATERIALS 装饰材料

The designers use calm black and gray to foil modern and fashionable atmosphere. The application of white TV wall stone and fresh color furniture reduce dark color heaviness and add vitality exclusive to the youth. One side of the wall in the dining room is covered with black mirror, by using the reflect feature of it to make lights shuttle in the interior, which not only extends space vision, but also promotes brightness of the residence.

设计师以沉稳的黑灰色系烘托现代时尚氛围，透过白色的电视墙石材运用、鲜艳跳色的家私配置，挥别暗色沉闷，增添专属年轻人的活力感。而在餐厅一侧墙面更以黑镜贴附，利用镜面的反射特性，让光线于室内穿梭，不仅延伸空间视觉，也提升居家的色调明亮度。

◎ **NATURAL LIGHTING** 自然采光

The designers take advantage of large area of window to bring in enough natural lights and brighten the public space in black and gray tone.

设计师善用大面开窗的采光优势，引入充足的自然光源，点亮以黑灰为主调的公共空间。

◎ **NATURAL COLORS** 自然色彩

The whole space sets black, white and gray as the main tones. Black focuses sight, white lifts brightness and gray deduces fashion. Fresh orange chair and soft ware inject young vitality into sedate atmosphere.

整体以黑白灰色调为主：黑色聚焦视线，白色提亮明度，灰色演绎时尚，通过鲜艳的橘色单椅、软件，为沉稳氛围注入年轻活力。

大道至简
SIMPLIFYING PROFOUND PHILOSOPHY

自然风 / NATURAL STYLE

THOUGHTFUL AND FUNCTIONAL RESIDENCE
贴心机能宅

◎ DESIGN CONCEPT 设计理念

In order to realize the dream of a home, the owner gets away from the hustle and bustle metropolitan area to the nearby city and is willing to be a commuter of high-speed railway. He would like the most reassuring and comfortable living environment for the family rather than a house shelter. The designers realize the dream concept of the owner by using "round" throughout the whole field from porch, corridor wall to main lamp of the dining room. Strong round vocabularies endow the space with a unique personality, grind the pattern edges of the layout and soften the vision with neat and warm atmosphere in light and bright tone.

为了圆一个家的梦想，屋主远离喧嚣的都会地区，选择来到邻近城市筑巢，而之所以愿意成为高铁通勤族，不只在于想有个遮风避雨的屋子，更深切盼望的，是给家人一个最安心、最舒适的生活环境。设计师实体化屋主的圆梦理念，以"圆"为造型贯穿整个场域，从玄关一路延伸至廊道壁面、餐厅主灯，强烈的圆弧语汇让空间有了独特个性，却也磨去格局的制式棱角，在清浅明亮的调性铺陈中，柔化视觉赋予利落不失温暖的氛围表现。

Project name ｜ Price Fragrance
项目名称 ｜ 太子馥

Design company ｜ Interplay Space Design Studio
设计公司 ｜ 演拓空间室内设计

Designers ｜ Ted, Ycy
设 计 师 ｜ 张德良、殷崇渊

Photographer ｜ KyleYu Photo Studio / Yuwei Huang
摄 影 师 ｜ 游宏祥摄影工作室/黄钰威

Location ｜ Taoyuan, Taiwan
项目地点 ｜ 台湾桃园

Area ｜135m²
项目面积 ｜ 135m²

Main materials ｜ laminate hard plastic sheet, paint wood veneer, glass, fog quartz brick, super durable wood floor, cultural stone, custom-made mirror stainless steel lamps, large picture output, etc.
主要材料 ｜ 美耐板、涂装木皮、玻璃、雾面石英砖、超耐磨木地板、铁件、文化石、订制镜面不锈钢灯具、大图输出等

◎ DECORATIVE MATERIALS 装饰材料

The designers thoughtfully plan kitchen ware functions which fit human engineering and use equipments such as smoke curtain and exhaust fan to strengthen the function of discharging smoke in the kitchen, which solves the problem of lampblack. Considering the location in humid area, the designers implicitly hide the dehumidifier into the ceiling, which breaks the existing position of dehumidifier.

设计师贴心规划符合人体工学的厨具机能，并以防烟垂壁、抽风机等设备，加强厨房排烟功能，解决开放式厨房易产生油烟的问题。同时考虑到本案坐落在气候潮湿的地区，设计师匠心独运，将除湿设备以吊隐式手法藏入天花板，颠覆既定摆放除湿机的做法。

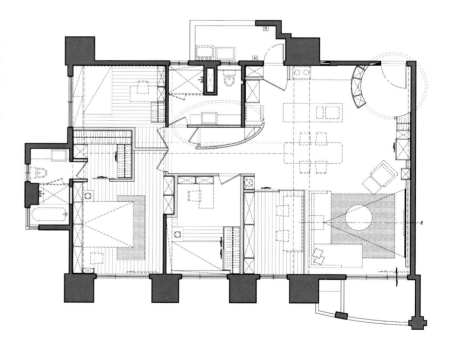

◎ SPACE PLANNING 空间规划

The designers connect public areas to reduce unnecessary partition to string functions of living room, dining room and kitchen, which makes the couple who loves cooking interact with the family when cooking. The study behind living room uses glass as the partition to create transparent vision, which makes the space independent and connective with public areas.

设计师将公共领域串联，减少不必要的隔间，无阻碍串联客、餐、厨机能，让喜爱下厨的屋主夫妇，能在料理的同时与家人亲密互动。同时，将位在客厅后方的书房，以玻璃为隔间创造通透视野，让空间既独立又能与公共领域串联。

◎ NATURAL COLORS 自然色彩

The whole space uses light colors and neat lines to outline the style. The living room matches fresh cyan pillows with gray cloth furniture to create warm living atmosphere. Warm wood furnishings and black and white stripe carpet create harmonious visual feelings. The concise and neat space designs present concise aesthetics to the extreme.

全室以清淡色彩、利落线条勾勒风格样貌，客厅用清新的青色靠枕搭配灰色调的布质家具，营造温馨居家的氛围；温润的木色摆件及黑白条纹地毯，创立和谐的视觉感受，简洁利落的空间设计，展现简约美感的极致。

TRANSPARENT BEAUTIFUL VILLA
透天美墅

◎ DESIGN CONCEPT 设计理念

This transparent villa of four floors is located in the edge of the city. Because of the configuration and house orientation, every floor doesn't have better lighting and view except the top floor. So the designer makes full use of the convex layer and focuses on connecting other interior spaces with the whole space and creating atmosphere.

Given that traditional transparent residence makes spaces over independent and separate because of floor height, the designer tries to use design methods and materials to effectively link vertical spaces and strengthen the overall and consistent space atmosphere on the premise of not changing the layout. So the designer creates an oak dying wood decorative wall from the first floor to the ceiling of the forth floor, collocating with vertical straight and thin line metal stair handrail and teak solid wood armrest, which makes the whole space connect together efficiently and vertically.

这是一栋位于城市边缘的4F连栋透天别墅，因为小区配置与房屋坐向，除了顶楼的屋突楼层之外，其他楼面并未享有良好的采光与视野。因此设计师除了将屋凸层的优势尽力发挥之外，便着力于其他内部空间的整体空间串联与氛围塑造。

有鉴于传统透天住宅因为楼层而将空间分割的过于独立与疏离，设计师企图在不变动隔局的前提下，利用设计手法与素材来将垂直空间有效地串连并强化整体一致性的空间氛围。为此，设计师创造出一道从1F一直往上延伸至4F天花板的橡木染色木皮装饰墙面，搭配垂直细线条的金属楼梯栏杆与柚木实木扶手，让整栋空间有效率垂直的串连起来。

Project name ｜ Freedom
项目名称｜森自在

Design company ｜ Vattier Design
设计公司｜瓦第设计

Designer ｜ Guohuan Huang
设 计 师｜黄国桓

Photographer ｜ Max Chung
摄 影 师｜钟崴至

Location ｜ Taoyuan, Taiwan
项目地点｜台湾桃园

Area ｜ 330m²
项目面积｜330m²

Main materials ｜ wood veneer, leather, gray lens, wood floor, tile, marble, iron part, etc.
主要材料｜木皮板、皮革、灰镜、木地板、磁砖、大理石、铁件等

◎ DECORATIVE MATERIALS 装饰材料

The whole space uses white oak steel brush wood veneer dyed in dark color as the axis, matching with metal, stone, glass and leather, which makes every function produce different material dialogues in different needs. At the same time, the choice of furniture adheres to the principle of dialogues with the space. So concise modeling, appropriate scale and harmonious color become the standard of the collocations.

整个空间以染深色白橡木钢刷木皮为主轴，搭配金属、石材、玻璃与皮革等，让各个机能不同空间在不同使用的需求下产生不同的材质对话。同时，在家具对象的选择上秉持着与空间对话的原则，因此造形简约、尺度合宜、色彩协调等成了选择搭配的基准。

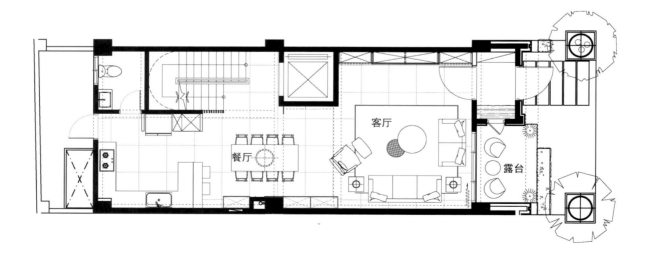

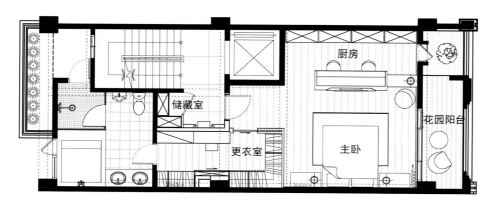

◎ SPACE PLANNING 空间规划

The designer makes full use of horizontal and vertical interactive rhythm of lines, which makes the height and width of the space extend visually. At the same time, the main axis of lines links each field which echoes with each other, creating a low-key, harmonious and profound living space.

充分利用了线条的水平垂直交互律动，让空间的高度、宽度获得视觉上的延展。同时，以线条为主轴，串联起各个场域，彼此独立又相互呼应，层层迭迭间创造出了一个既低调和谐却又隽永深刻的居住空间。

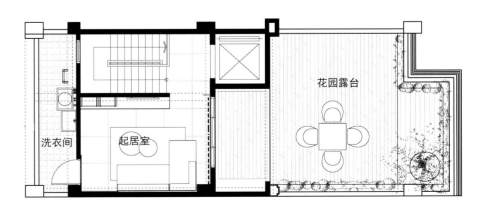

大道至简
SIMPLIFYING PROFOUND PHILOSOPHY

自然风 / NATURAL STYLE

GENTLE BREEZE WITHOUT FORM
风和·无相

◎ DESIGN CONCEPT 设计理念

This project obscures the definition of indoor and outdoor space, interprets sanative residence by natural materials and creates a leisure and comfortable living atmosphere for the owner. The inside and outside areas integrate, using virtual-real light and shadow extensions and open and close actions of materials to break the boundary of the space, which makes the interior space extend infinitely to outside. Penetrable glass material narrates design attempts to bring lights into interior. The water mould wall which can calm you down brings outside indifference into interior. The plain wood materials make the space return to nature. Jumping out of the layering veins, there is only light and shadow. The designs are hidden in details. The plain and clean wood matches with simple colors, which endows the space with a distinct gradation.

本案模糊室内外空间界定，以自然材质诠释疗愈住宅，为屋主打造休闲舒适的住家氛围。内外区域相互交融，借由虚实交映的光影延伸与介质的开阖动作，打破空间界线，将室内空间无限延伸至户外。穿透的玻璃材质，述说引光入室的设计企图。让人心情沉静的清水模墙，将属于户外的淡定，纳入室内，质朴的原木材质让空间回归自然本质。跳脱层次脉络，独留光影。在光影交融中，纯粹的材质与比例丰富了空间层次，宁静且感动。设计隐藏于细节中，在素净的木质与朴质的色彩搭配下，让空间更显层次分明。

Design company ｜ PJDESIGN
设计公司 ｜ 品桢空间设计有限公司

Designer ｜ CHEN YING-HSIN
设 计 师 ｜ 陈膺信

Photographer ｜ LDK Photography Studio
摄 影 师 ｜ 利德凯国际空间摄影团队

Location ｜ Taipei, Taiwan
项目地点 ｜ 台湾台北

Area ｜ 132m²
项目面积 ｜ 132m²

Main materials ｜ iron part, glass, water mould, Laos fragrant juniper, wood grain tile, oak smoked sea island floor, underfloor heating, etc.
主要材料 ｜ 铁件、玻璃、清水模、寮国香桧、木纹砖、橡木烟熏海岛型地板、地暖等

◎ NATURAL COLORS 自然色彩

The space uses wood color and gray to set low-key and texture living atmosphere, collocating with white ceiling and bar, black leather sofa, cloth carpet and floral soft coverage dining chairs, which creates sedate and restrained sense of quality for the dining room and living room. Removing the dark calmness, the designer adds warm color brightness. Two unique chairs in passionate red and clean transparent color complement with each other in the study. Orange bed and light wood book cabinet make the study different from other spaces with a jumpy space rhythm.

空间以木色、灰色奠定低调质感的居住氛围，搭以白色的天花、吧台，黑色皮质沙发、布艺地毯和花色软包的餐椅，为餐客厅空间营造出沉稳内敛的品质感。除去深色的稳重，设计师也加入暖色的明丽，书房两把造型独特的椅子，红色的热情，透明色的明净，互为补充，橙色的软塌和浅木色的书柜，让书房有别于其他空间，显示其跳动的空间旋律。

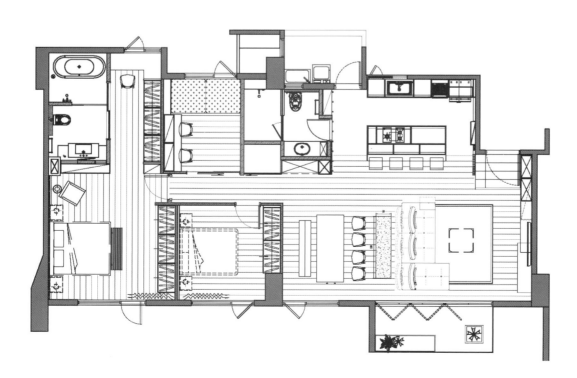

◎ NATURAL LIGHTING 自然采光

Sparse and mottled sunshine slowly sparkles in the balcony. Natural and warm lights always give people comfortable and soft feeling. To bring maximum lights into interior, the space uses large area of French window to bring natural lights into interior to form a virtual-real impression. Big terrace design, spacious space and white space make lights communicate freely in the space, which creates a humanistic living environment beyond space.

稀疏斑驳的阳光缓缓洒下阳台，自然和煦的光芒总给人舒适柔和的感觉。为了最大化引入室外光线，空间采用大面积的落地窗设计，让光线自然而然导入室内，形成虚实相生的印象。大露台的设计和宽敞的空间、设计留白，让光线在空间自由流动对话，创造属于超越空间界限的人文居住环境。

大道至简
SIMPLIFYING PROFOUND PHILOSOPHY

自然风 / NATURAL STYLE

CONCISE LIVING RESIDENCE
简约生活居所

◎ DESIGN CONCEPT　设计理念

Located in Taoyuan, this Dai Residence uses wood as the main tone, collocating with Zen-like furniture, which makes every corner full of concise wood taste.It uses the owner's favorite Japanese style as the main axis.The designers start with exquisite details and texture space designs, control five senses experiences produced by light and shadow, material texture and temperature, time and sequence transformation and create a strong Japanese space character.

位于桃园的戴宅，以木质作为居所主调，搭配具有禅风意味的家具，让角落每一处都充满木质简约味道。该案以业主喜爱的日式风格为主轴，透过讲究细节与质感的空间设计着手规划，掌控光影、材质肌理与温度、时序转移产生的五感体验，打造浓厚的日式空间性格。

Project name ｜ Taoyuan Dai Residence
项目名称 ｜ 桃园——戴宅

Design company ｜ REZO WORK CO., LTD
设计公司 ｜ 日作空间设计

Designers ｜ Ray Huang, Chiang Kuan-I, Chen Chuan-Hao
设 计 师 ｜ 黄世光、江冠逸、陈隽澔

Photographer ｜ KyleYu Photo Studio
摄 影 师 ｜ 游宏祥摄影工作室

Location ｜ Taoyuan, Taiwan
项目地点 ｜ 台湾桃园

Area ｜ 198m²
项目面积 ｜ 198m²

Main materials ｜ cypress, oak, cedar, iron part, tatami, kieselguhr, imitated wood grain water mould, etc.
主要材料 ｜ 桧木、橡木、香杉、铁件、榻榻米、硅藻土、仿木纹清水模等

◎ NATURAL LIGHTING 自然采光

Large area of window in the living room brings lights into interior, collocating with warm light color wall and carpet, which creates a light and bright recreational space. The designers set a window in the kitchen which is convenient for sunshine to pour into interior. Full of natural light and fragrance of wood, happiness arises spontaneously, which also reduces forcing feeling from the top space.

客厅中的大面积开窗引光入室，搭配温润的浅色系墙面及地毯，创造轻盈敞亮的休闲空间。设计师在厨房中也设计了方便阳光倾泻的开窗，盈满自然光线与木质的芬芳，幸福感油然而生，同时也减弱了顶部空间带来的逼仄感。

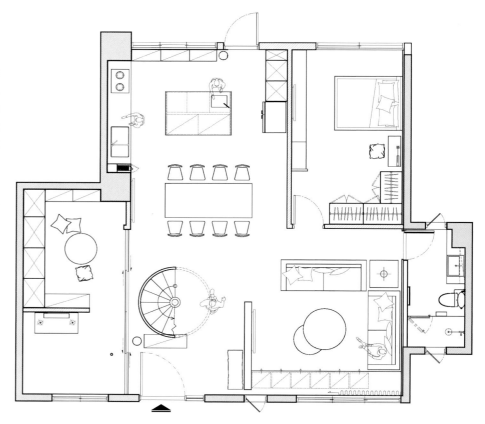

◎ SPACE PLANNING 空间规划

Spiral stairs hover above with strong dynamic beauty, and you can have different visual experience in different angles. At the same time using spiral stairs as "interval point", the designers use life kinetonema naturally to divide living room and dining room, which maintains their respective independence and creates an open layout.

螺旋状楼梯盘旋其上，具有强烈的动态美，从不同的角度可得到不同的视觉体验。同时以螺旋状楼梯为"间隔点"，运用生活动线自然地切分出客厅、餐厅等区域，保持各自空间独立的同时打造开放的格局。

◎ NATURAL COLORS 自然色彩

The furniture continues the warm texture of wood, collocating with big window to bring in natural lights, which forms a casual and pure living space with restrained gray and warm beige. The bedroom uses the same sedate wood tone, soft and comfortable bedding and orderly interval window to create a comfortable and cozy living atmosphere.

　　家具延续木质的温润质感，搭配大开窗引进的自然光线，与内敛的灰色、温暖的米色一起，组成了一个随性纯粹的生活空间。卧室同样采用沉稳的木质色调，与柔软舒适的床品、间隔有序的开窗，共同营造了舒适惬意的居家氛围。

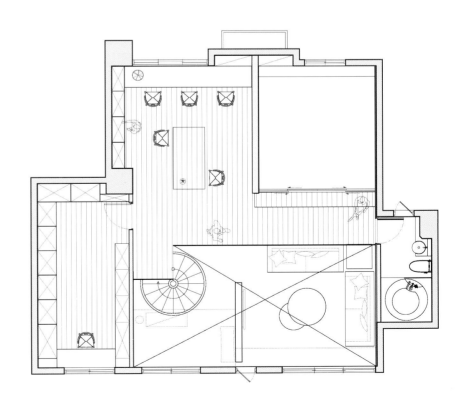

大道至简
SIMPLIFYING PROFOUND PHILOSOPHY

自然风 / NATURAL STYLE

HOLIDAY LANDSCAPE RESIDENCE
坐拥度假景观宅

◎ DESIGN CONCEPT 设计理念

This project is located in mountain top residential area with elevators in New Taipei City with mountain and sea scenery, and is a place for the owner to take a holiday.

The designers highlight providing residents field where they can relax and comfort their bodies and minds. In addition to satisfy functions of general life, it stresses communicating with the exterior environment including leading in flowing air, comfortable sunshine, capacious and comfortable space scale and holiday atmosphere. Large areas of open windows bring in maximum lights. The changeable lights and shadows endow the space with lively colors. The perfect mount top residence has views about all the mountain and sea landscape. Opening any window can bring you holiday feelings.

本案位于新北市山顶电梯集合住宅，坐拥山海大景，是屋主假日休闲度假之居所。

设计师重视提供居住者在身心上能放松安适的场域，除一般生活机能上的满足，更着重与室外环境的对话，包括引入流通的空气，舒适的阳光，宽敞舒服的空间尺度，以及着重度假般的疗愈氛围。大面积的开窗设计，将光线最大化的引入，在一朝一夕的光影变化中，赋予空间更多的灵动色彩。而绝佳的山顶住宅位置，坐拥一切山海景色，任意打开一扇窗，都将你带入度假感受。

Project name | Mid-Mount Owner
项目名称 | 半山主人

Design company | TWADESIGN
设计公司 | 传十设计

Designers | Tiangui Xu, Wenxin Lee
设 计 师 | 许天贵、李文心

Location | New Taipei City, Taiwan
项目地点 | 台湾新北

Area | 133m²
项目面积 | 133m²

Main materials | wood floor, tile, paint glass, iron part, latex paint, etc.
主要材料 | 木地板、磁砖、烤漆玻璃、铁件、乳胶漆等

◎ SPACE PLANNING 空间规划

The public areas such as living room, dining room and rest room adopt open space planning, which has a horizontal continuation of window scenery and extends the vision. The tall rest room leaning on the window becomes a landscape residence for owners to relax. There is intentionally a glass partition wall between the master bedroom and the rest room, which makes the view of bedroom extend to the living room, dining room and rest room and brings in sunshine and scenery.

将客厅、餐厅、卧榻区与厨房等公领域，采取无隔间之开放平面计划，让水平窗景连续，视觉延伸。倚窗架高的卧榻区，成为主人慵懒放松的观景处所。在主卧室与卧榻区之隔间墙，刻意留出一道玻璃隔间，让卧室的视线延伸到客、餐厅与卧榻区，同时也引入日光与景色。

◎ NATURAL COLORS 自然色彩

The modeling style which doesn't intend to shape strong style mainly uses clean and comfortable earth tone. The dark warm wood floor and beige wall set the main tone of the space, collocating with horizontal cabinet combined with iron part and wood, which reflects more spectacular scenery. The natural log dining table makes a strong contrast with the modern and concise paint glass. The tall rest room leaning on the window becomes a landscape residence for owners to relax.

不刻意形塑强烈的造型风格，以清爽闲适的大地色系为主，深色温暖的木质地坪与米色墙面为空间基调，搭配铁件与木作组合的水平柜体，呈现轻盈无压的内敛调性。餐厅区的深色烤漆玻璃墙，随着四季晨昏景象变化，反射映照出更为壮阔之景致，而充满自然气息的原木餐桌，更与现代简炼的烤玻形成强烈对比。一旁规划倚窗的架高卧榻区，成为一处慵懒放松的观景处所。

◎ NATURAL LIGHTING 自然采光

The master bedroom is equipped with concise and horizontal cabinet and bed modeling, collocating with indirect lighting and designs which don't fall to the ground, which makes the space lighter. There is intentionally a glass partition wall between the hall and the rest room, which makes the view of bedroom extend to the outsides and brings in beautiful sunshine and scenery. The bathroom continues transparent partition which endows the bath space with sufficient lights and sceneries, relating ubiquitous comfort and coziness.

　　主卧室以简洁水平延伸的柜体与床体造型，搭配间接照明与不落地设计，使量体更显轻盈。而与厅区卧榻区之间，则刻意留出一道玻璃隔间，使卧室视线可延伸至外，同时导入美好的日光及景色。浴室延续透明隔间的设定，让沐浴空间内盈满光线与景致，诉说无处不在的自适惬意！

大道至简
SIMPLIFYING PROFOUND PHILOSOPHY

自然风 / NATURAL STYLE

INHERITING FAMILY RESIDENCE, BEGINNING A SANATIVE NEW LIFE

传承世代的家族居所 展开疗愈的新生活

◎ DESIGN CONCEPT 设计理念

This is a modern residence in the original land and a gift from the father to the son, declaring the continuation of the family from generation to generation. Space from the construction, the exterior landscape to indoor decorations is planned by the designer. The building appearance has an inclined housetop extremely full of design sense. The linear and neat modeling becomes highlight of the street. Under the planning of the designer, it not only has bright manner through abundant lights, but also integrates profound aesthetic connotations with unique tastes, creating a unique landscape villa.

The architectural design has day lighting in three sides, each of which can introduce a lot of natural lights, making the whole space bright and full of vitality. It has a total of four floors. Each floor of the building has a height of more than 3.6 meters. Considering its future convenience, it also sets an elevator inside. The stair with vertical kinetonema adopts light iron part structure collocating with warm wood floor, which is the focus of the space and increases air flowing and penetration of it. You can feel the breath of the architecture.

这是一幢自地自建现代宅，是父亲留给儿子的礼物，宣告着家庭世代的延续。空间从建筑、外部景观到室内皆全权交由设计师规划，建物外观呈现斜屋顶样貌、极富设计感，以流线利落的造型构成街景亮点。在设计师的规划之下，不仅透过丰沛采光串起明亮气场，更融入深厚的美学底蕴与独到品味，打造出独具一格的别墅景观。

建筑设计三面采光，每个面都可以引进大量的自然光，让整个空间明亮且充满生命力。总共四层楼，每层楼的楼高均有三米六以上之高度，考虑到其未来的方便性，也在屋内设置了电梯。垂直动线的楼梯采用轻盈的铁件结构搭配上温暖的木地板踏面，不仅是空间中的焦点，也增加空间的空气流动性与透光性，可感觉到建筑的呼吸。

Project name ｜ Display
项目名称 ｜ 展

Design company ｜ Herzu Interior Design
设计公司 ｜ 禾筑国际设计

Designer ｜ Tam
设 计 师 ｜ 谭淑静

Location ｜ Taipei, Taiwan
项目地点 ｜ 台湾台北

Area ｜ 470m²
项目面积 ｜ 470m²

Main materials ｜ wood veneer, stone, stoving varnish metal, polish quartz brick, super durable wood floor, cultural stone, glass, etc.
主要材料 ｜ 木皮、石材、烤漆金属、抛光石英砖、超耐磨木地板、文化石、玻璃等

◎ SPACE PLANNING AND NATURAL LIGHTING 空间规划、自然采光

The designer firstly brings a lot of natural lights into the whole building, creating vitality and texture, then makes changes of the modeling light troughs and droplight through the extension of the inclined ceiling lines in the 4.2 meters tall space on the first floor, which links visual layering between the living room and dining room and indicates that the gathering family can feel different atmospheres created by lamps as if under the tree. The place under the stairs to the second floor matches iron part with wood floor, which not only hides the cover plate of the storage room, but also has the function of free and flexible bed. The vertical and transparent linearity makes outside beautiful easy to get. According to the couple's habits, two desk areas are set and combined with low TV cabinet, which increases the interaction between the couple in the space!

设计师首先替整幢楼引入大量自然光，营造朝气与质感，并在一楼四米二的挑高空间里，透过斜天花的线条延伸，在造型灯槽与吊灯之间做出变化，串起客厅、餐厅之间的视觉层次，也暗示着相聚的家人就像是在树下一样，其灯具也可随着不同的使用方式营造不同的氛围。通往二楼的楼梯下方，则利用铁件搭配温润木地板，不仅隐藏储水室盖板，也兼具自由灵巧的卧榻机能，借由垂直通透的线性，让楼梯有如空气般的轻盈，型构整幢楼的主视觉。卧房内则透过窗户导入自然光，使室内可轻易接受到室外美好景致，同时因应夫妻两人的使用习惯，设置两处书桌区，并让电视矮柜结合书桌机能，增加两人在空间中的互动性！

◎ BRINGING SCENERY INTO HOUSE 引景入室

Inclined housetop created in accordance with laws and regulations removes general inertia technique and is turned into the highlight of the building ending. Large area of windows in the vertical staircase kinetonema increases areas of natural light illumination. It uses shadow of sunlight after pouring into the wall through the stairs to form the most beautiful reflection. Large area of French windows in the audiovisual room links balcony with indoor spaces as a whole. Weakening the relationship between the inside and outside of the building, the outdoor greenery naturally becomes side view of the interior, which obscures boundaries of each space and keeps its future. The inherited home is not just a home, but also infinite possibilities.

依照法规而成的建筑斜屋顶，除去一般惯性手法，使之转换成整栋建筑收尾的亮点。大面开窗在垂直楼梯动线增加了自然光照射面积，利用阳光经过楼梯洒落在墙面的阴影形成最美丽的倒影。视听室的大面落地窗，将露台及室内空间整体串联。将建筑内外的关系弱化，户外的绿意自然形成室内的端景，模糊各空间的边界，保留其未来性。传承的家不只是家，还有无限的可能性。

DESIGN CONCEPT 设计理念
SPACE PLANNING 空间规划
DECORATIVE MATERIALS 装饰材料
NATURAL LIGHTING 自然采光
NATURAL COLORS 自然色彩
INTERIOR VIRESCENCE 室内绿化
BRINGING SCENERY INTO HOUSE 引景入室

LIGHT
LUXURIOUS STYLE

轻奢风

interpreting concise tone, tasting light
luxurious charm

雕刻简约情调，品味轻奢魅力

大道至简
SIMPLIFYING PROFOUND PHILOSOPHY

轻奢风 / LIGHT LUXURIOUS STYLE

CREATING CLASSICAL METROPOLIS RESIDENCE
缔造古典都会居所

◎ DESIGN CONCEPT 设计理念

Located in central area of Xinzhuang District in New Taipei City, this project has top location, convenient traffic and perfect life functions and is an ideal living environment in metropolis. The integral space uses low-key and concise classical vocabularies, collocating with restrained and texture materials and sedate and elegant colors, which forms a magnificent atmosphere for the space. With the concept of private hostel as the starting point, it creates a welcoming and treating atmosphere, supplemented with art works to present uniqueness of the space, which creates an artistic and humanistic elegant residence.

本案位于新北市新庄区正都心精华地段，为拥有顶级地段、便利交通及完善生活机能的新建个案，是都会生活中理想的居住环境。整体空间运用低调而洗练的古典语汇，搭配内敛质感的材质与沉稳优雅的色调，形塑空间大气感。并以私人招待所的概念为出发点，营造出迎宾宴客的情境氛围，辅以艺术品点缀空间的独特性，缔造富有艺术人文的雍雅居所。

Project name ｜ Tudors Show Flat
项目名称 ｜ 都锋苑样品屋

Design company ｜ Trendy International Interior
设计公司 ｜ 动象国际室内装修有限公司

Designer ｜ David Tan
设 计 师 ｜ 谭精忠

Cooperating designers ｜ Xinhui Lai, Qingrong Lin
参与设计 ｜ 赖欣慧、林青蓉

Location ｜ New Taipei City, Taiwan
项目地点 ｜ 台湾新北

Area ｜ 307m²
项目面积 ｜ 307m²

Main materials ｜ spray lacquer, titanium, iron part, steel brush wood veneer, wall cloth, carbonized wood floor, stone, yarn glass, gray lens, leather, etc.
主要材料 ｜ 喷漆、镀钛、铁件、钢刷木皮、壁布、碳化木地板、石材、夹纱玻璃、灰镜、皮革等

◎ DECORATIVE MATERIALS AND NATURAL COLORS 装饰材料、自然色彩

The porch matches steel brush wood veneer with dark leather wallboard modeling. So does the ceiling. The consistent space deposits the soul and spreads a prelude. The space uses dark tone to create sense of mystery, supplemented with plating titanium, which balances the heavy color through refraction of materials. Contemporary art paintings endow the porch with restrained and texture visual atmosphere. The wallboard modeling combines with collecting space. The inside cabinet has functions of collecting caps, clothes and shoes and also has absolute practicality.

玄关以钢刷木皮搭配深色皮革的壁板造型，并在天花板运用相同的设计语汇，一致性的空间使人沉淀心灵，展开进入的序曲。空间以深色调来营造神秘感并辅以镀钛金属来做点缀，透过质材的折射平衡了较浓重的色调，并置放当代艺术画作，让玄关弥漫内敛质感的视觉氛围。壁板造型与收纳空间结合，柜内另藏有兼具衣帽与鞋子的收纳功能，也具备了绝对的实用性。

◎ SPACE PLANNING 空间规划

Entering from the porch, you can see the open living room and dining room. The spacious space presents the mansion manner with low-key and concise steel brush wood veneer wall vocabulary throughout the whole space. What's more, the modeling bucket box and red wine display cabinet continue the original design techniques, in addition with leather, iron part and titanium, which elaborates texture and layering of the space and presents a magnificent and elegant manner. The main wall of the living room uses travertine to foil the painting "native rhythm" by contemporary artist Ceng Yongning, creating a unique visual taste. The entire space creates an atmosphere of private club. Exquisite furniture, collected wine and well-chosen art works echo with each other, which presents uniqueness and aesthetics of the space and conveys restrained and sedate life attitude which is worth to taste carefully.

The open light food kitchen enlarges the space scale of the living room and dining room and continues their design vocabularies. In addition with the island which maintains the light food kitchen functions, it plans display space for art plates to present its uniqueness under the foil of yarn glass and iron part, which not only integrates the light food kitchen and dining room into a whole, but also makes the light food kitchen enrich atmosphere of the whole space.

　　由玄关进入，映入眼帘的是客厅与餐厅的开放式空间，宽敞的空间展现豪宅的气度。以低调洗炼的钢刷木皮壁板语汇来贯穿整个空间，另在造型斗框、红酒展示柜上不仅延续原有的设计手法，也加入皮革、铁件及镀钛金属来作细节上的处理，铺陈空间的质感与层次，呈现大气雍雅的气势。客厅主墙以洞石来衬托当代艺术家曾雍宁的画作"原生的律动"，创造出独有的视觉韵味。整体空间营造成私人会所宴客的情境，并以精致的家俬、典藏的美酒、万中选一的艺术品三者相互呼应，来呈现空间的独特感与美学氛围，进而传达内敛沉稳却值得细品的生活态度。

　　开放式的轻食厨房更开拓了客餐厅的空间尺度，在视觉上也延续客餐厅的设计语汇，除了有岛台维持轻食厨房的机能外，另也规划艺术杯盘展示空间，在夹纱玻璃与铁件的衬托下让展示品更显出其特有性，不仅让轻食厨房与餐厅融为一体，也让轻食厨房丰富了整体空间的氛围。

◎ NATURAL LIGHTING 自然采光

The master bedroom design sets a comfortable and magnificent tone. Main walls use wall cloth and leather as the basement and use transparent and soft lights inside the iron part ray cutting totem to echo with outdoor gray-blue sky, which presents elegance and uniqueness of it. At the same time, the master bathroom uses stone to deduce high texture space, collocating with top sanitary equipments and contemporary art paintings, which is as if in an exclusive restaurant. Enjoying the bath with open view from French windows and bringing lights of dawn into indoor, you can fully relax your body and mind and enjoy yourself.

主卧室的设计以舒适且大气的基调来呈现，在重点墙面皆以壁布及皮革为基底并运用铁件雷切割图腾内透柔和的灯光，与室外灰蓝的天空相应，呈现出主卧室的雅致与独特。同时主浴室以石材演译高质感的空间并搭配顶级的卫浴设备与当代艺术画作，宛如置身高级饭店。在享受泡澡的当下，伴随落地窗辽望的开阔视野，让晨曦之光引入到室内，使身心充分的放松与享受。

大道至简 / SIMPLIFYING PROFOUND PHILOSOPHY

轻奢风 / LIGHT LUXURIOUS STYLE

TEMPERATURE OF TIME
时间的温度

◎ DESIGN CONCEPT 设计理念

With the pursuit of innovation and uniqueness and the pursuit of personality and fashion, art life becomes the mainstream and personality household design becomes a trend. Most people have their own aesthetic ideas in the living environment of highlighting their own styles itself and expect unique shapes and complete functions to bring the home more surprises.
This project is one of the humanistic and luxurious villa show flat in Chongqing Yuanyang golf, conveying temperature of the space by highlighting humanistic and luxurious flavors, which endows essences of space design into this project. The overall style is pure, fresh and refined, stressing quality and comfort and convenience of living life. From materials to modeling, layout to colors, every space connects with each other and highlights its own uniqueness. Even the detail is full of exquisite joy!

追求创新与独特，追求个性与时尚，艺术生活成为主流，个性家居设计成为一种潮流。大多数人在追求突显自我风格的家居环境下，拥有自己的美学观点，希望个性化定制，期待独特的造型、全面的功能性给家带入更多的惊喜。

本案是设计师在重庆远洋高尔夫做的其中一套人文奢华风的别墅样板房，以突出人文奢华的韵调来传递空间的温度，很好地将空间设计的精髓展现在本案中。整体格调清新雅致，注重品质，突出居住生活的舒适性与便利性。从材质到造型，从布局到色彩，每个空间都彼此相连又突出自己的独特性。细品之处，皆有精喜！

Project name ｜ Chongqing Yuanyang Golf 180 House Type
项目名称 ｜ 重庆远洋高尔夫180户型

Design company ｜ TINE FUN Interior Planning Ltd
设计公司 ｜ 天坊室内计划

Designer ｜ Qingping Zhang
设 计 师 ｜ 张清平

Location ｜ Chongqing
项目地点 ｜ 重庆

Area ｜ 406m²
项目面积 ｜ 406m²

Main materials ｜ marble, wood, stainless steel, etc.
主要材料 ｜ 大理石、木质、不锈钢等

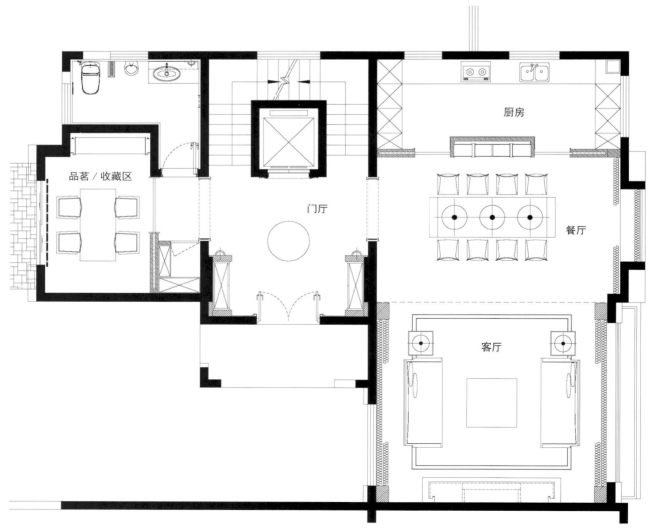

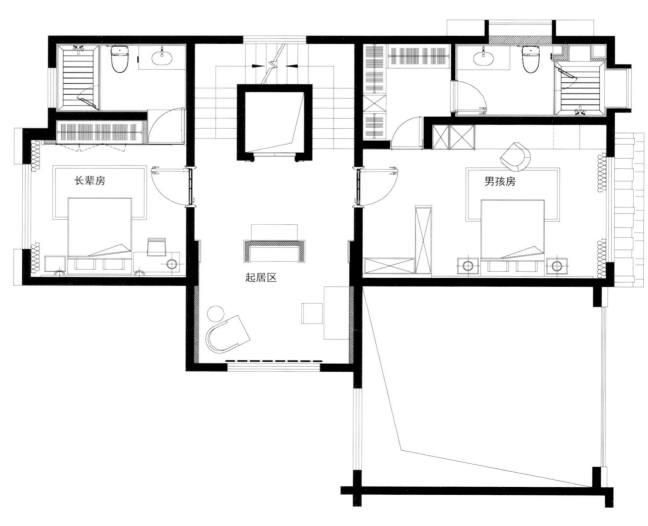

◎ NATURAL LIGHTING 自然采光

With broad floor height and double open windows, the lights penetrate into interior along with the white curtain, making the entire living room bathed in sunlight, which is leisure and peaceful so that we can fully enjoy energy and temperature given by magical nature. This is a special emphasis from the designer to the owner, which is the owner's favorite healthy and active lifestyle.

开阔的层高，大开窗的双面设置，光线沿着白色的帘幔层层渗透进来，让整个客厅都可沐浴在阳光之下，闲适安然，充分享受神奇的大自然赐予我们的能量与温度。这便是设计师特别强调，传达给业主，希望业主喜欢的一种健康积极的生活方式。

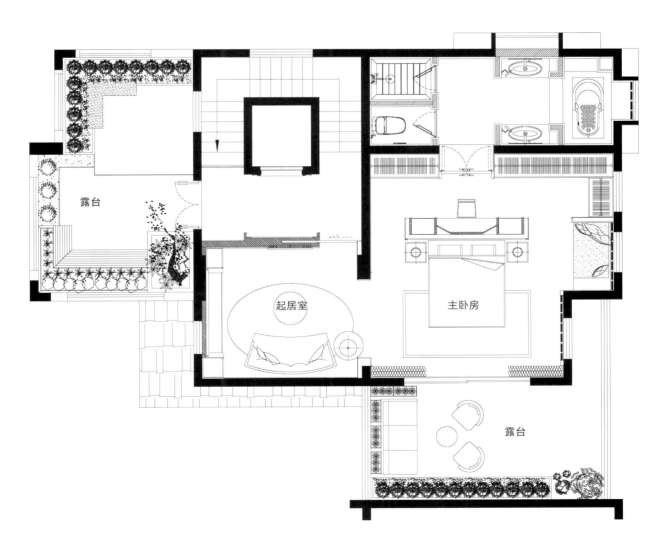

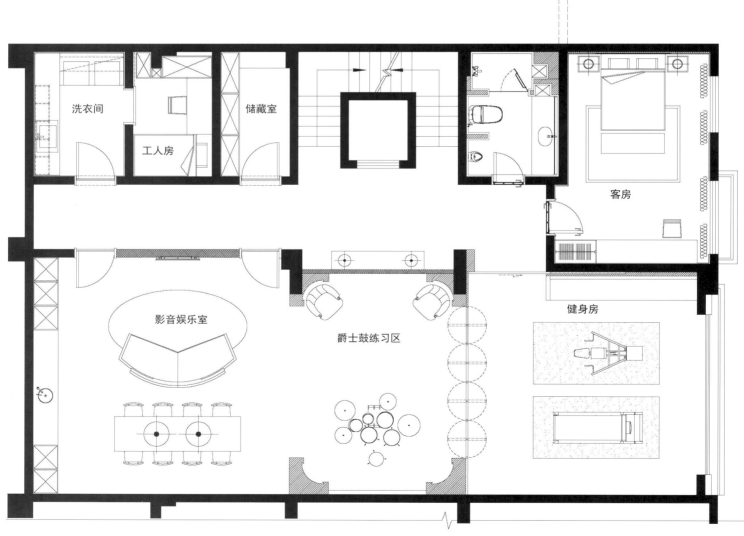

◎ NATURAL COLORS 自然色彩

Cleanness is the most salubrious tone of the whole project. Plain white, concise gray, bright black and elegant brown and lake green are pure elaboration in each surface without a little impurity, bringing the owner the purest visual taste.

　　干净，是整体案例给我们最清爽的色调感受。质朴的灰、简约的白、以及高光的黑和优雅的褐色及湖绿色，都是纯洁的铺陈在每一个平面上，不掺和一点点杂质，带给主人最纯粹的视觉品味。

大道至简 | SIMPLIFYING PROFOUND PHILOSOPHY

轻奢风 / LIGHT LUXURIOUS STYLE

ENSEMBLE OF WOOD AND STONE BECAUSE OF A WILLING HEART
木石重奏 金石为开

◎ DESIGN CONCEPT 设计理念

The villa of five floors has unique and open manner with large areas. In order to manifest the exquisiteness and dignify of the mansion, the designer uses wood and stone to present delicate vertical texture. It specially uses the hostess's name as inspiring element, integrates with classic and elegant orchid modeling and creates a distinct and unique interior atmosphere by personalized designs.

The whole space is open with unique modeling, which creates a magnificent momentum. Warm woods are used in the top surface of the living room, which creates a natural atmosphere, collocating with titanium indirect lighting design and crystal droplight, which promotes texture of the space. The favorite gray U-shape leather sofa is spacious and comfortable. Regardless of gathering or being alone, it has ineffable geniality and vitality and endows the home with quality life.

本案为5层楼的别墅，拥有大坪数独有的开阔气势，为凸显大宅的精致轩昂，设计师以木石为媒材，精彩呈现细腻的立面质感。特别撷取女主人姓名作为灵感元素，融入古典优雅的兰花造型点缀，以个性化设计创造出鲜明独特的室内风貌。

整体空间感开阔，造型独特，形塑出磅礴的气势。温润的木皮铺陈延展整个客厅顶面，营造出自然氛围，搭配镀钛间接照明设计和水晶吊灯，提升空间质感。最爱灰色U型皮革大沙发，宽大舒软，无论聚会或者独处，都有着不可言喻的亲切与活力，赋予家更多的品质生活。

Project name ｜ Xue Residence
项目名称 ｜ 薛宅

Design company ｜ Yuli Interior Design
设计公司 ｜ 由里室内设计

Designer ｜ Irene Fu
设 计 师 ｜ 傅琼慧

Photographer ｜ Woodman
摄 影 师 ｜ 大头人

Location ｜ Tainan, Taiwan
项目地点 ｜ 台湾台南

Area ｜ 347m²
项目面积 ｜ 347m²

Main materials ｜ stone, metal titanium, wood veneer, spray lacquer, wallpaper, glass, bright mirror, etc.
主要材料 ｜ 石材、金属镀钛、木皮、喷漆、壁纸、玻璃、明镜等

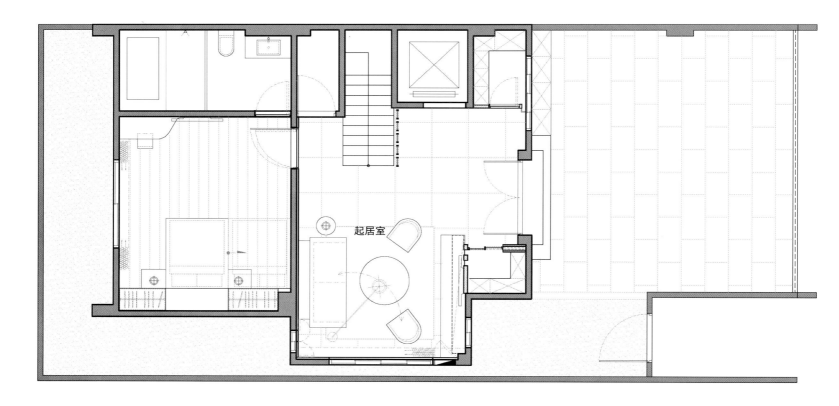

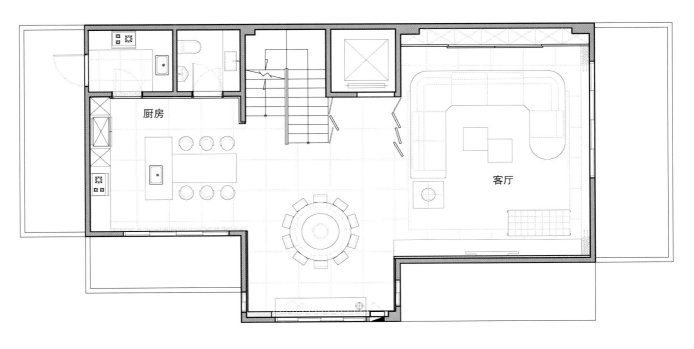

◎ DECORATIVE MATERIALS 装饰材料

The well-chosen layering texture marbles make the facade form natural images with waves and water, creating a dynamic visual effect and bringing a magnificent momentum. The ceiling with wood elements presents warm and plain natural flavors and manifests a perfect collocation with the rough and sedate stone. Moderate titanium metal materials interpret low-key and luxurious mansion manner by modern design techniques.

　　精心挑选纹理层层堆栈的大理石，让立面形成海浪、流水的自然意象，创造出律动的视觉效果，带来磅礴恢弘的气势。以木质元素做搭接的天花板，延展出温润朴质的自然气息，与粗犷沉稳的石材展现绝妙的搭配。适度加入镀钛金属质材，以现代的设计手法，演绎低调奢华的大宅风貌。

◎ NATURAL LIGHTING 自然采光

The living room uses titanium glass partition to make natural lights reflect beams of light, which creates bright and spacious vision and injects vigorous vitality into the space. The entire glass balcony design combines with the outside courtyard landscape, creating a leisure and relaxing atmosphere.

客厅区域以镀钛玻璃隔屏，将援引入室的自然光折射出一道道光束，创造出明亮宽敞的视野，为空间注入蓬勃生命力。而整面的玻璃阳台设计，结合室外的庭园造景，围塑出休闲舒压的氛围。

◎ SPACE PLANNING 空间规划

Half tall marble TV wall keeps penetrable sights from both sides and above, which forms circulatory rectangular-ambulatory-plane kinetonema and closely links living room, sleeping area and relaxing area. In the bedroom, the bright display bookcase is the entrance and barrier. The transparent design brings rich layering and depth of field. The living room uses iron part and yarn glass to create a partition effect, combining with orchid modeling carvings, which forms a beautiful barrier.

半高中岛大理石电视墙，保留两侧和上方的穿透视线，形成循环的回字动线，将客厅、卧眠区、卧榻区紧密串连。卧房内，以明亮的展示书柜兼作入口屏障，通透的设计，带来丰富的层次和景深。而起居室处，利用铁件和夹纱玻璃创造出隔屏效果，结合兰花造型雕刻，形成一道美丽的屏障。

大道至简 / SIMPLIFYING PROFOUND PHILOSOPHY

轻奢风 / LIGHT LUXURIOUS STYLE

EXTREMELY DEDUCING HOUSEHOLD ART LIFE
极致演绎家居艺术生活

◎ DESIGN CONCEPT 设计理念

Located in the entrance of Cultural Second Road and Zhongxiao Road in Linkou, this project is the new target of Shintoshin, with convenient transportation and perfect life functions. The show flat uses low-key and concise design vocabularies, collocating restrained and texture materials with colors, which creates a magnificent space atmosphere and highlights the design key of this project which is a performance stage with extreme coexistence of exquisite life and household art.

Entering from the entrance into the living room and dining room, you can feel capaciousness and manner throughout the spaces. About 10 meters wide open space vision presents magnificent mansion manner. Floors of public spaces are covered with gray-black marbles, which forms a strong contrast with white paint wall and foils calmness and restraint of the space. Regular cutting and dyed dark steel brush wood veneer on the wall divide the space region by soft technique and present tension of the space.

本案位于林口文化二路及忠孝路口，为林口新都心指标。便利交通、完善生活机能的新建个案。样品屋以低调而洗练的设计语汇，内敛而富质感的材质与色调搭配，形塑空间大气感，彰显出本案的设计重点：精致生活与家居艺术极致共容的展演舞台。

由入口进入客、餐厅区时，入内即能感受贯穿客厅与餐厅之开放式空间的宽敞与气度，面宽近10米宽开阔的空间视野，呈现豪宅气势。公共空间地坪铺设灰黑色的大理石石材，与白色喷漆壁面形成强烈的对比，进而衬托出空间的沉稳与内敛。墙面规律性的分割与染深色的钢刷木皮，不仅将空间区域性以软性手法加以分界，更加以铺陈空间的张力。

Project name | Linkou World Chief Show Flat
项目名称 | 林口世界首席样品屋

Design company | Trendy International Interior
设计公司 | 动象国际室内装修有限公司

Designer | David Tan
设 计 师 | 谭精忠

Cooperating designers | Zirong Chen, Yuzhen Xu
参与设计 | 陈姿蓉、许玉臻

Location | New Taipei City, Taiwan
项目地点 | 台湾新北

Area | 172m²
项目面积 | 172m²

Main materials | solid wood steel brush wood veneer, spray lacquer, titanium, wall cloth, oak wood floor, gray mesh stone, carved white, leather, yarn glass, gray lens, etc.
主要材料 | 实木钢刷木皮板、喷漆、镀钛、壁布、橡木地板、灰网石、雕刻白、皮革、夹纱玻璃、灰镜等

◎ DECORATIVE MATERIALS AND SPACE PLANNING 装饰材料、空间规划

The ceiling of the living room sets white as the tone and creates sense of layering by linear framework stacks. The top of the ceiling is embedded with meticulous hand-made shell. Exquisite detail treatments create a unique visual taste. The living room combines with the reading area, which enlarges visual effects of the space. The display cabinet in the reading area combines titanium framework with dark wood steel brush wood veneer to preset high texture, which becomes side view of the living room.

Cabinets in the living room and dining room adopt concealed treatments, continue the cutting effects of the walls, make the entire space lines clean by hiding behind the wall and endow the space with high collecting function. White paint wall hanging painting art works foil the elegant and tasty atmosphere in the space and strengthen the effect after resident combines with art.

客厅区以白色为基调的天花板，运用线性框架堆栈上升营造出层次感。在天花最高处，以细致手工贝壳镶嵌在天花上方，精致的细节处理，创造出独有的视觉韵味。同时客厅结合阅读区，不仅拉大空间的视觉效果，在阅读区展示柜体的处理，以镀钛金属框架结合深色钢刷木皮以精品家具呈现出高质感展示柜，更成为客厅区域端景。

客、餐厅区域的柜体均使用暗柜方式处理，延续墙面分割效果的同时，隐身在墙面后方的柜体，使整体空间线条干净外，更有高机能的空间收纳功能。素白的喷漆墙面吊饰画作艺术品，衬托出空间里优雅与品味，强化了住宅与艺术结合后的加分效果。

◎ NATURAL COLORS 自然色彩

The yarn door pieces of the dining room side view wall belong to kitchen area. The white film glass material used for ceiling lighting brightens the kitchen. Under the foil of super white paint glass, the well-equipped white kitchen ware makes the kitchen space cleaner and neater.

The master bedroom maintains the consistent fashion, with advanced hotel configuration as the orientation and open toilet and bedroom area as configuration concept, which enlarges its entire space feeling. On material selection, it continues dark steel brush wood veneer of the living room to foil carved white stone basement in the bathroom, which presents sedate and restrained texture. The headboard of the master bedroom is flexible to open and close. It remains the original function of window and combines titanium frame embedded with manual shells with flexible sliding door pieces, which makes the entire wall more delicate and fashionable. The changing room with complete functions is exquisite and connects with the bathroom. The closet adopts functional design method, combining with gray lens and dark wood veneer, which extends sense of space.

餐厅端景墙的夹纱门片为厨房区域，天花板照明设计白膜玻璃材质规划，明亮了厨房空间。设备齐全的白色厨具在墙面上超白烤漆玻璃的衬托下，让厨房空间更显干净利落。

主卧室的设计维持一贯时尚，以高级饭店配置为取向，将厕所与卧房区使用开放式效果作为配置理念。不仅拉大主卧房的整体空间感受，在材质选用上，延续客厅的深色刚刷木皮衬上厕所白色为基底的雕刻白石材，延续沉稳与内敛的质感。主卧床头板，可弹性开关，维持原有窗户使用机能外，并以镀钛框镶嵌手工贝壳与活动式拉门片结合，让墙面整体更具精致与时尚。更衣室功能完整且精美，与浴室相连接。衣柜设计采用功能性的设计方式，并以灰镜与深色木皮作结合，延伸空间感。

大道至简
SIMPLIFYING PROFOUND PHILOSOPHY

轻奢风 / LIGHT LUXURIOUS STYLE

DETAIL AND COOL ABSTRACTION
细致与冷抽象

◎ DESIGN CONCEPT 设计理念

Pursuing harmonious coexistence between man and environment has been a civilized human thought core. In the process of urbanization, however, though people have bigger and bigger spaces, they are in a tense environment. Under such a demand, space tends to be clean and concise and the design theme will be integrated into thought and life of the owner and bring more convenience and comfort. In this project, the designer injects Italian style into this 400-square-meter mansion. He uses Minimalism as philosophy and emphasizes on the importance of space to ration, line, geometry and proportion. He regards broader field as a container, paints it with furniture and low chroma tone to make clean colors stretch completely on vision through natural lights and to return to space tension which belongs to heaven, earth and own.

追求人与环境的和谐共处，一直是文明人类的思考核心；然而，在城市化的过程中，拥有的空间越来越大，但人却处于一种紧张的环境，在这样的需求之下，空间会趋于干净简洁，设计主题也会融入屋主的思想与生活，并带来更多的便利与舒适性。在本案中，设计师为这间400平方米的宅邸，注入属于意大利的设计风采，他以简约主义（Minimalism）作为圭臬，强调空间之于理性、线条、几何、比例的重要性。他更将开阔的场域视为容器，以家私及低彩度的色调涂绘，让干净的色系透过自然光线，在视觉上完全舒展开来，重返属于天地与自我的空间张力。

Design company ｜ VERY SPACE INTERNATIONAL
设计公司 ｜ 咏义设计股份有限公司

Designers ｜ Louis Liou, Eaten Huang, Hsiao-Chieh Chou
设 计 师 ｜ 刘荣禄、黄沂腾、周筱婕

Photographer ｜ Kyle Yu
摄 影 师 ｜ 游宏祥

Location ｜ Taoyuan, Taiwan
项目地点 ｜ 台湾桃园

Area ｜ 400m²
项目面积 ｜ 400m²

Main materials ｜ dark gray quartz brick, smoky oak, oak dyed gray, pine stone, tawny glass, iron part, etc.
主要材料 ｜ 深灰石英砖、烟熏橡木、橡木染灰、松柏石、茶镜、铁件等

◎ DECORATIVE MATERIALS 装饰材料

Starting from the capacious living room, sofa and display cabinet as if abstract sculpture works and ceiling which keeps sense of cool abstraction and deliberately deepens the layering changes create rich light and shadow changes to reflect in the abstract glass porch made of dichroic glass. A lot of material elements and delicate arrangements echoing with each other and mutual penetrable and reflective rich space visual perception create an abstract aesthetic space full of rich sense of layering.

从宽敞的客厅开始，如抽象雕塑作品般的沙发与展示柜，乃至刻意加深创造层次变化，且同样保有冷抽象设计感的天花板，而其创造的丰富的光线变化，亦反射在由双色玻璃组成的抽象玻璃门廊中。透过诸多彼此呼应的材质元素与细腻安排，相互穿透、映像的丰富空间视觉感，创造出具有丰富层次感的抽象美学空间。

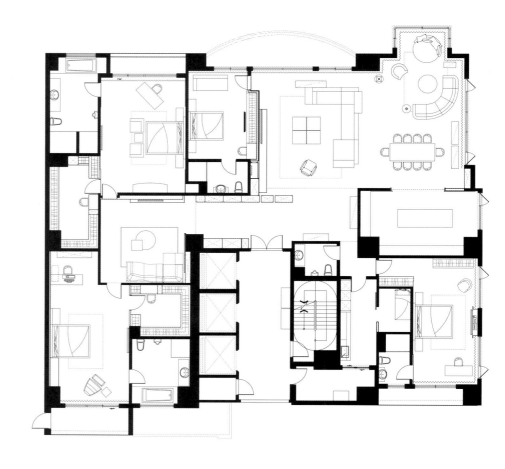

◎ NATURAL COLORS 自然色彩

The space uses dark brown as the base, with gray and warm white in the top to echo it. The designer chooses various material properties and colors, endows them with slight red, slight green, dark or light delicate changes and hides more profound emotional textures in them. Being with yourself is a very important moment, gently put yourself into the warm mansion from trivial real life to experience washing and adjustment of mind and to experience flowing amorous feeling and happiness of life again.

空间看似皆为深褐色为底，上方则以灰暖白与之相应，设计师仍透过各种材质属性与选色，在其中混入了偏红、偏绿或深或浅的细腻变化，并在其中藏入了更为深刻的情感肌理。与自己相处，是一个很重要的时刻，轻轻地把自己从琐碎的现实生活中，融进温和的宅第里，经历一番心情的洗涤与调整，重新感受生活的流动风情与美好。

大道至简
SIMPLIFYING PROFOUND PHILOSOPHY

轻奢风 / LIGHT LUXURIOUS STYLE

GORGEOUS CASTLE
华丽城堡

◎ DESIGN CONCEPT 设计理念

Located in the prosperous area of city center, this residence contains 14th, 15th and 16th floors of a 24-floor high-end residential building. The low-key owner buys three floors, where he can enjoy convenient life functions and safety hotel-style property management services and hopes that the designer can create the feelings of modern single-family villa residence. So without destroying beam structure of the building, how to effectively connect three floors to maintain unity of space design atmosphere and create a horizontal and vertical space of single-family villa with open and transparent senses becomes the biggest challenge to the designer.

The style of this project continues gorgeous neoclassical building origin of the overall building, absorbs aura of nature and pays attention to combining style with practicability to make "home" become a veritable castle.

这是一个位于市中心繁华区域，总高 24 楼的高端住宅大楼内 14、15、16 楼的一个住宅单位。低调的业主买下 3 个楼层，除了能享受方便的生活机能及安全良好的饭店式物业管理服务外，还希望设计师能够创造出独栋现代别墅住宅的感觉。因此在不破坏大楼梁柱结构下，如何有效地串连三个楼层维持空间设计氛围的统一性，并创造出独栋别墅的水平垂直空间，开放、穿透感成了设计师最大的挑战。

本案风格设计沿袭整体项目一贯华丽气派的新古典建筑血统，同时吸收自然界的灵气，注重风格与实用结合，让"家"成为一座名副其实的城堡。

Project name ｜ Zhongyue Imperial Residence
项目名称 ｜ 中悦皇苑

Design company ｜ Vattier Design
设计公司 ｜ 瓦第设计

Designer ｜ Guohuan Huang
设 计 师 ｜ 黄国桓

Photographer ｜ Moooten Studio
摄 影 师 ｜ 墨田工作室

Location ｜ Jubei, Taiwan
项目地点 ｜ 台湾竹北

Area ｜ 830m²
项目面积 ｜ 830m²

Main materials ｜ wood veneer, marble, iron part, etc.
主要材料 ｜ 木皮板、大理石、铁件等

◎ DECORATIVE MATERIALS AND SPACE PLANNING 装饰材料、空间规划

Firstly the designer needs to find the most appropriate position in the space to define the position of stair which strings all vertical spaces. The unchangeable beam structure of the super high building becomes a relatively difficult challenge. After careful thinking and simulation, the designer creates a 10-meter tall space and a modern black steel stair, which effectively and interestingly links the whole space. In addition with vertical connection, the designer further seeks horizontal link of the space and uses distribution of function configurations and glass opening penetrable technique to maximize vertical and horizontal senses of public areas of the whole living space. There is an 8-meter high white stone wall in the tall space, which creates virtual and real echoing dialogue with the penetrable glass opening and makes obvious contrast with the nearby black staircase, highlighting this black staircase with modern structure lines.

设计师首先要做的是在空间内，最合宜的位置，定义出串联整个垂直空间的楼梯位置。梁柱结构无法被改变的超高层大楼变成了相对困难的挑战。经过了缜密的思考与模拟，设计师创造出了一个高10米的挑空区域，一座极富现代感的黑色钢梯，有效且有趣地联系了整个空间。除了垂直的串联之外，设计师更进一步寻求空间的水平链接，利用空间机能配置的分配与玻璃开口的穿透手法，让整体居住空间的公共领域垂直与水平开阔感达到最大化，并在这个挑高空间置入一道高8米的白色石材墙面，与穿透的玻璃开口产生虚与实的呼应对话，也与紧邻的黑色楼梯产生明显对比，让这颇富现代结构线条的黑色钢梯更具存在感。

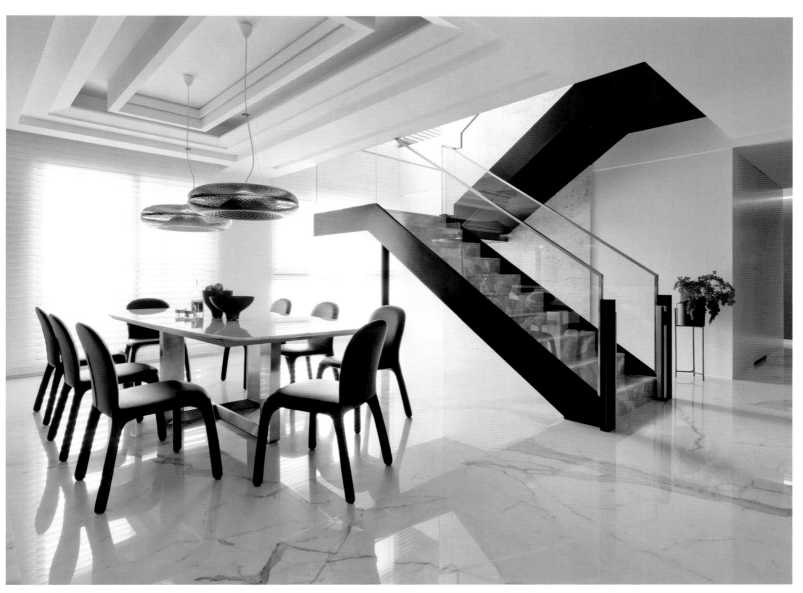

◎ BRINGING SCENERY INTO HOUSE 引景入室

The designer installs a set of high and low modeling lamps with sculptural senses in the top of the tall space to endow residents with visions in different angels and layering no matter day or light when moving or staying horizontally and vertically.

Looking up or looking down occasionally, space fun of different floors happen here. People living here often forget that they are in the high building, which creates a villa space in the building.

设计师在挑空区域的顶部安装了一组高低错落、空间雕塑感十足的造型灯具，让在这个房子生活的住户无论白天或夜晚，在水平垂直移动或停留驻足时，都能够拥有不同角度与层次的视野。

时而仰望或时而俯视，不同楼层的空间趣味在这里四处发生，在这里生活的人们常忘了身处于高楼中，创造出这个大楼中的别墅空间。

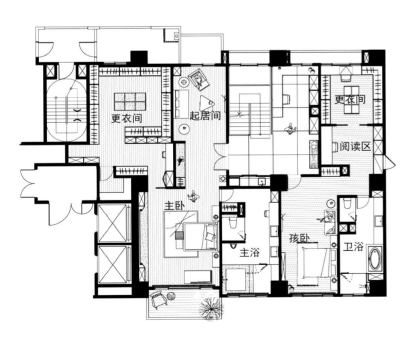

THE EXTREME STYLE
极致格调

◎ DESIGN CONCEPT 设计理念

This is a show flat of a real estate in the eleventh floor of the building. The designer chooses modern light French style to create a high-quality and comfortable high-end residence. The biggest challenge of this project is to present the high-end texture in some degrees and effectively leave a deep impression on clients within limited budgets.

Considering visual confusion and Fengshui troubles because the entrance faces directly with the door of the subaltern room, the designer sets a metal screen at the entrance. The metal laser cutting paper-cut screen creates a visual focus at the entrance, effectively changes the direction of the entrance kinetonema and clearly defines the porch area, which makes the space sequence full of layering functions and more complete and clearer.

这是个位于大楼11F的地产开发商样板房项目，设计师选择了现代轻法式风格的诠释，打造一处集品质与舒适于一体的高端住宅空间。本案的最大挑战来自于有限的预算限制内，必须呈现相当程度的高端质感并有效地让客户留下深刻的印象。

鉴于入口大门正对次卧房间门造成视觉上的纷乱，设计师便于入口处定义出一道金属屏风，这道金属雷射切割剪纸屏风不仅创造出入口视觉焦点，更有效地改变了入口动线方向，清楚地定义出玄关区域，让空间序列更有层次机能更为完整清楚。

Project name ｜ Mountain Hanlin Rich Residence
项目名称 ｜ 山璞翰林富苑

Design company ｜ Vattier Design
设计公司 ｜ 瓦第设计

Designer ｜ Guohuan Huang
设 计 师 ｜ 黄国桓

Photographer ｜ Max Chung
摄 影 师 ｜ 钟崴至

Location ｜ Taoyuan, Taiwan
项目地点 ｜ 台湾桃园

Area ｜ 330m²
项目面积 ｜ 330m²

Main materials ｜ wood veneer board, wallpaper, tawny glass, marble, iron part, etc.
主要材料 ｜ 木皮板、壁纸、茶镜、大理石、铁件等

◎ DECORATIVE MATERIALS 装饰材料

Floors of the original public area stress bright and elegant. On the premise of no changes, the designer chooses dark gray coffee Italian water dyed wood veneer to collocate with dark geometric line wallpaper, which makes the space more stable and sedate. Some parts are interspersed with gold elements such as Spanish antique chair, dining droplight and desk lamps, which emphasizes dignified texture.

原公共空间地坪材料着重明亮与素雅，设计师在不更动的前提下，选用了深灰咖啡色意大利水染木皮，搭配深色几何线条壁纸，让空间多些安定沉稳。再局部点缀金色元素，如西班牙古董单椅、餐吊灯与台灯等物件，强调尊荣质感。

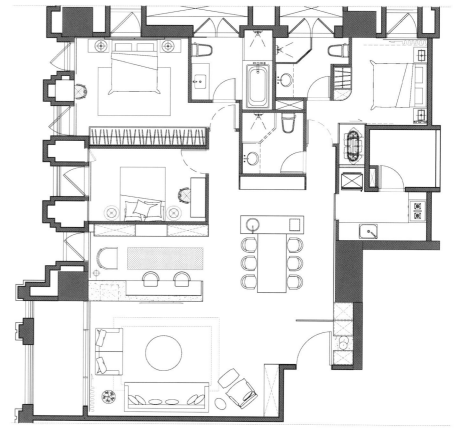

◎ NATURAL COLORS 自然色彩

Because the scale of the bedroom is not big, the designer wants to use salubrious color collocation and smooth texture to make the space spacious and comfortable. Therefore, the cabinet uses stoving varnish to reduce wood grains and avoid visual burden caused by too many lines. The salubrious powder-blue and pink wallpaper and leather provide a stabilizing effect.

卧室部分因为尺度不太大，设计师希望利用清爽的配色与平顺的质感让空间相对感觉宽敞与舒适。因此，在柜体的部分选择了烤漆涂装处理减少木纹的表现，避免过多的线条造成视觉的负担，而清爽的粉蓝、粉红色系壁纸与皮革使用则提供了安定情绪的效果。

大道至筒
SIMPLIFYING PROFOUND PHILOSOPHY

轻奢风/ LIGHT LUXURIOUS STYLE

THE IMPERIAL RESIDENCE
帝之苑

◎ DESIGN CONCEPT 设计理念

Human-oriented life philosophy is able to meet functional needs of daily life, and most importantly is able to become the communication field to condense family life. Whether it is night lighting or day decoration, it can endow the interior space with more exquisite texture atmosphere. This project is a five-floor transparent residence. The designers stress the planning of the space layers. The capacious space matches with natural and comfortable decorations with flowing air and natural lights. With sun rising and moon falling and a slowly breeze, large areas of windows communicate with the outside scenery, which maximally endows the space with changeable scenes.

以人为本的生活哲学，要能满足生活起居的机能需求，以及最重要的，要成为凝聚家庭生活的交流场域，无论是夜间点亮、或日间装饰都能赋予室内空间更细腻的质感氛围。此案为五层楼的通透住宅，设计师注重空间层次的规划，宽敞的空间搭配自然舒适的装饰，同时引入流动的空气、自然的光线，日升月落，微风徐徐，通过大面积的窗户对话室外，最大化地赋予空间变化的光景。

Project name | Zhang Residence
项目名称 | 张公馆

Design company | Tianching Spatial Design
设计公司 | 天境空间设计

Designers | Fu Han Tsai, Tzu Han Hung
设 计 师 | 蔡馥韩、江怡婵

Photographer | Junjie Liu
摄 影 师 | 刘俊杰

Location | Taichung City, Taiwan
项目地点 | 台湾台中

Area | 297m²
项目面积 | 297m²

Main materials | wood veneer, paint, imported wallpaper, marble, framed cloth, leather, iron part, etc.
主要材料 | 木皮、油漆、进口壁纸、大理石、裱布、皮革、铁件等

◎ SPACE PLANNING 空间规划

The five-floor space planning is well-organized. The double choices of elevator and stairs link every floor. The first floor is mainly a garage and is designed with bathroom, kitchen and island bar, which is convenient to go out. The second floor is public area which is a common place for the family gathering and for them to condense emotions. The living room and dining room is a open combination, which creates a capacious and comfortable sense. The warm island kitchen with water channel is planed as the second arranging area. When friends come, they don't need to wait aside and can feel involved in the bar. The third and forth floor are bedrooms and rest areas. The master bedroom and guests room are in the third floor. The second master bedroom and the daughter's room are in the forth floor. The fifth floor has a roof terrace and a study where the owner can read and work quietly.

五层空间规划井井有序，电梯和楼梯的双重选择连接各层空间。一层以车库为主，同时设计卫生间、厨房、中岛吧台等，起到方便出行的作用。二层是公领域，公领域是家人团聚共同使用的空间，也是一家人感情凝聚的地方。客厅及餐厅为开放式的组合，创造出宽敞舒适感，温馨的中岛厨房有水槽被规划为第二料理区，朋友来不用在另一个空间干等，可以在中岛吧台有参与的感觉。三楼和四楼以卧房休息区为主，主人房和客房在三楼，第二主人房和女儿房分布在四楼。五楼有着屋顶楼台，还有可安静在此看书办公的书房。

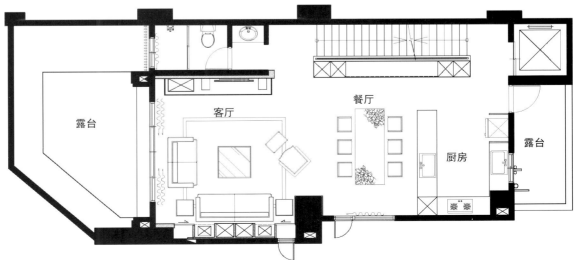

◎ DECORATIVE MATERIALS 装饰材料

The sofa background wall of the living room has the functions of collecting and displaying. The iron part presents light and concise furnishing frames. The light iron part, warm wood floor and light and elegant imported wallpaper, collocating with texture furnishings, add beauty to the space. the TV wall stone uses opposite grains to outline the features of the stone, which manifests distinct line beauty of ink painting.

客厅沙发背墙兼具收纳及展示的功能，利用铁件表现出轻盈简易的陈设框体，铁件的轻薄感加上木皮温润的层板，壁面为浅淡雅致的进口壁纸，再搭配上有质感的饰品，更为空间加分，电视墙石材以对纹的方式勾划出此石材的特色，显现鲜明波墨画的线条美感。

The sideboard in the dining room uses retro handles to collocate with sandblasting wood veneer and concise long octagonal linear schindylesis modeling, which presents ingenious technique of the designers and pure aesthetics and exquisite hand crafts quality. The waist has modern hexagonal iron part laser cutting. Part of the penetration presents virtual and real senses, which makes the space penetrate and extend, releases oppressive feelings and makes lights sprinkle on the stairs.

餐厅餐边柜使用复古把手搭配喷砂木皮及简约的长型八角型线条沟缝造型设计，呈现设计师巧思的手法，展现出纯粹美感与精致手工制作的品质，腰带部分有现代感六角形铁件雷射切割，部分呈现虚与实的穿透感，让空间穿透、延伸，减少压迫感，也可以让光线洒下楼梯。

◎ NATURAL LIGHTING 自然采光

With light gauze and curtains and a gentle breeze, the living room and master bedroom adopt large areas of French window, which naturally brings in vernal sunshine and beautiful scenery. Floor lamps and wall lamps at bedside naturally integrate the interior and outside lights. At the same time, there is a small window at the head of the bed, which makes the interior lighting more sufficient and can be modification.

轻纱帷幔，清风习习，客厅和主卧采用大面积的落地窗，自然导入和煦的日光和明媚的景色，落地灯和床头壁灯，使内外光线自然融合在一起，同时房间床头开有一扇小窗，不仅让空间采光更加充足，同时也起到修饰作用。

LINE EXPRESSION AND FRAME CARRIER
线语·框载

◎ DESIGN CONCEPT 设计理念

Original, plain and intentionally secluded gray background properly foils randomly displayed furniture of two floors. The walls intentionally keep trowel coating texture without modification, in addition with the projector lamps, which is like strokes that apply forces on the paper, delivers uniqueness of the furniture and creates casual definition of the home.

The concept of "frame" implements design thoughts from outside to inside. The tabular stereo view frame of the architectural facade dialyzes energy of light and shadow. The first floor presents three parallel areas in L-shape. The frame lines of the ceiling edge form a visual carrier. The rich turn of layering leads the conversation among materials, space and environment and then reveals deep reflection between conflict and harmony.

原质、无华、刻意隐退的灰阶背景，称职地衬托两个楼层间错落陈列的家具。墙面刻意保留未经修饰的镘刀涂层手作质感，加上聚焦灯光穿针引线，类似施力于素描纸上构思"线"的笔触，传递家具的独一无二，打造家的随性定义。

"框"的概念，落实由外而内的设计思维，建筑外观扁平的立体景框，透析光影能量。一楼呈现L型串接的三个平行区块，利用天花板边缘的框线勾勒，形塑视觉载体，层次丰富的转折间，引导素材与空间、环境的对话，进而揭示冲突与和谐之间的深度省思。

Design company ｜ YYDG INTERIOR DESIGN
设计公司 ｜ 源原设计

Designers ｜ Peny Hsieh, Calvin Tsai
设 计 师 ｜ 谢佩娟、蔡智勇

Location ｜ Taipei, Taiwan
项目地点 ｜ 台湾台北

Area ｜ 198m²
项目面积 ｜ 198m²

Main materials ｜ tile, spray lacquer, iron part, etc.
主要材料 ｜ 地砖、喷漆、铁件等

◎ DECORATIVE MATERIALS AND NATURAL LIGHTING 装饰材料、自然采光

Taking full use of good lighting of the first floor, large area of French window brings lights into interior. Iron part elements with neat lines and lamp frames are used in furniture. The functions are set in blocks, presenting a unique texture. The floor and wall have sedate and magnificent grains, which acts as a foil and complement with all the furniture.

充分利用一楼的优良采光,以大面积落地窗援引光线,并以线条利落的铁件元素与灯光框设家具本体,将功能以区块方式设置,展现独特质感。

地坪与墙面皆以色调沉稳大气的纹路呈现,达到衬托作用,与所有家具相得益彰。

◎ NATURAL COLORS 自然色彩

The collocation and arrangement of furniture and software not only are extensions of interior designs, but also reflect life aesthetic taste of the owner. On the basis of gray tone, the project adds sedate black and jumping orange, supplemented by elegant line modeling, which presents fashion and magnificence in leather texture.

家具、软件的搭配与布置是室内设计的延伸,更是体现屋主对于生活美学的品味。整体灰调的基础上,加入沉稳的黑色、跳跃的橙色,并辅以优雅的线条造型,在皮革的质感中,体会时尚大气。

大道至简
SIMPLIFYING PROFOUND PHILOSOPHY

轻奢风/ LIGHT LUXURIOUS STYLE

ROMANCE AND ART
漫·艺

◎ DESIGN CONCEPT 设计理念

The space uses free and organic artistic paintings to eliminate the conflict of the entrance and create enough cushion space. The symmetric lines outline the layering. The fields are injected with art and French high quality culture and borrow separations and polymerizations in different times to make the continuation of the space. The clear visual attachment presents life aesthetics of space and time and echoes with colorful and ecological future forms. The green scenery of the balcony sets off the colorful layering of the artistic screen, which is independent and integrated between the inside and outside.

空间由艺术自由、有机的画面，消弭入口的冲突，促成余裕的缓冲空间，依凭对称线条，勾摹出重重轮廓。将场域中介植入艺术，载入法式精品文化，借由不同时段的分离与聚合，交迭出空间的延续性，清透的转载视觉感动，体现空间与时间的生活美学，呼应多彩与生态共生的未来形式。阳台绿景衬出艺术屏风的多彩层次，内外之间，独立却融合。

Project name | Huang Residence in Taipei
项目名称 | 台北黄宅

Design company | EXCLAIM UNITED
设计公司 | 京玺国际

Designer | Joy
设 计 师 | 周燕如

Location | Taipei, Taiwan
项目地点 | 台湾台北

Area | 330m²
项目面积 | 330m²

Main materials | imported tile, French imported wall cloth, Italian imported marble, old oak solid wood floor, etc.
主要材料 | 进口磁砖、法国进口壁布、意大利进口大理石、老橡木实木地板等

◎ SPACE PLANNING 空间规划

The areas of the flat space connect with each other and there is a clear distinction between primary and secondary. The public areas of living room and dining room are open and connective. Warm gray L-shaped sofa emits a sense of quality, which forms delightful contrasts with the art screen and art painting. The low-key and luxurious crystal droplight of the dining room and the handbag on the display cabinet form the visual focuses, which presents color beauty of female fashion. The master bedroom is equipped with master bathroom and cloakroom, which provides comfortable rest experience. The subaltern room is concise and warm. The gray bedside surface furnishings and wood chest of drawers present warm texture.

平层的空间各区域之间相互流动，主次分明。公领域的客厅与餐厅呈开放式相连，暖灰色L形长条沙发散发出的品质感与艺术屏风、艺术画相映成趣。餐厅中低调华丽的水晶吊灯和展示柜上的手提包凝聚视觉焦点，呈现女性时尚的色彩美感。主卧房中主卫、衣帽间配备完整，为主人提供舒适的休息体验，次卧房简洁温馨，灰色床头面饰和木色衣柜装饰出温润如玉的质感。

◎ NATURAL LIGHTING 自然采光

Through the symmetrical relationship, the space foils reasonable contexts of the kinetonema, leads in outside lights and leisurely uses them to organize the winding kinetonema, intermediary corridor and progressive and clear expressions. At the same time, the context thought is used to achieve the progress that the inside and outside forms continue linear axis of the space, which makes object views, lines, layering and colors fade in the sun lights and transforms artistic emotions and cultural symbolizations into perceptional interpretations of exploring visions.

空间透过对称关系，烘托动线间合理性的脉络，导入室外光线之际，从容的借以组织出委迤动线、中介过道以及递进交错的分明表情。同时以涵构思维来达成内外形式延续空间线性轴向的前进感，让物景、线条、层次与颜色在光线中都依次薄淡，借此将艺术指涉的情感和文化象征性，转以探寻视觉的感知诠释。

◎ NATURAL COLORS 自然色彩

The sharing of humanity and art extends its unique melody and rhythm. The tension of gray floor, orange chairs, colorful pictures and white beddings symbolize the transformation of space attribute and create rich and restrained conflict beauty, which is balanced and symmetric with corresponding material characteristics and creates a harmonious beautiful state.

人文与艺术的共享，延伸出独特自有的旋律与节奏。灰色系地坪的张力，橙色系的单椅、彩色系的挂画和白色的床品，象征空间属性的转换，共同营造丰富与内敛的冲突之美，平衡对称、且对应材质特色，制作出和谐靓境。

大道至简
SIMPLIFYING PROFOUND PHILOSOPHY

轻奢风 / LIGHT LUXURIOUS STYLE

ENJOYING RURAL AND LEISURE LIFE
享田园快意生活

◎ DESIGN CONCEPT 设计理念

With the advantage of large area, the designer doesn't intend to cover the existing beams and walls, keeps the original large sense of space and replaces partition walls by large area of window scene by using the layout relations between the balcony and living room, which brings green scenes into interior, narrows the distance between human and nature and creates a European living atmosphere pursued by the owner.

Considering owner's requirements of having two studies, some part of the kitchen is transformed into the second study while the original kitchen is handled with closed design to avoid lampblack of blowing out when cooking, which can hide the inside modern kitchen ware and makes the entire style unobtrusive.

In order to originally present foreign leisure and comfort, the designer sets light color as the main tone, integrates rural, American and European style elements in different areas and creates a romantic and elegant aesthetics in the space. In addition, imported furniture collocates with previous old objects, which makes nostalgic flavor pervade in classic style and creates leisure life as if in a Western manor.

拥有大坪数优势的此案，设计师不刻意包覆既有梁柱与壁面，保留住原始的大尺度空间感，并利用露台与客厅的格局关系，以大面窗景取代其隔间墙，让绿意光景无拘束穿梭入内，拉近人与自然的距离，营造出屋主向往的欧洲生活氛围。

考虑屋主有两间书房的需求，撷取部分厨房段落转化为第二间书房，并顺势以封闭式处理原先的厨房，不仅防止料理时的油烟散逸，亦能遮掩内部的现代感厨具，让整体的风格表现不显突兀。

而为原汁原味呈现国外的休闲惬意，设计师先以轻浅色为空间的主色调，再糅合乡村、美式与欧式等风格元素，于各区块分别表现，交织一室的浪漫优雅美学。此外，选用国外进口的家具摆饰，并融入屋主的旧有对象，让怀旧气息弥漫经典之上，打造宛如西方庄园的快意生活。

Project name | Lee Residence
项目名称 | 李宅

Design company | Yuli Interior Design
设计公司 | 由里室内设计

Designer | Irene Fu
设 计 师 | 傅琼慧

Photographer | Mr. Mo
摄 影 师 | 老莫

Location | Tainan, Taiwan
项目地点 | 台湾台南

Area | 307m^2
项目面积 | 307m^2

Main materials | grain brushing, solid wood, plate, marble, iron part, figured glass, retro mirror, antique finish, etc.
主要材料 | 钢刷木纹、实木板、线板、大理石、铁件、压花玻璃、复古镜、仿古漆等

◎ DECORATIVE MATERIALS 装饰材料

The designer wants to create a diverse and romantic luxurious space for the owner through delicate crafts. Classic furniture modeling sets in symmetric place, which creates pure and authentic flavors of European style. At the same time the ceiling is modeled by plates with gold paint, manifesting palace magnificence. The walls are covered with totem wallpaper and modeling fireplace, presenting an elegant ritual aesthetics.

设计师希望透过细致的手工作法，为屋主打造出一处多元浪漫的奢享空间。经典造型家具，对称性的位置摆放，营造纯正、地道的欧风情韵。同时天花以线板造型，并以金漆绘出宫廷般的大器；墙面则铺以图腾壁纸与造型壁炉，展现优雅的仪式美学。

◎ SPACE PLANNING 空间规划

In order to jump out of the existing TV wall impression, it uses glass to let through free visions, brings the outside green into interior and reaches the state of unifying human and nature. At the same time, the ceiling is embedded with mirror to stretch visual height of dining room. The entrances to study and kitchen are designed with glass grille modeling doors, presenting classic American charms. The original area of kitchen is reduced to make a study, which meets owner's needs of having two studies.

为了跳脱既定的电视墙印象，采用玻璃让视觉穿透无碍，并顺势引入窗外绿意，达到人与自然合一的境界。同时，天花嵌入镜面拉伸餐厅处的视觉高度，两侧分别通往书房与厨房的入口，则以玻璃格子造型门，呈现经典美式风味。将厨房的原先尺度缩减，空间转为书房用途，满足屋主对两间书房的需求。

大道至简 | SIMPLIFYING PROFOUND PHILOSOPHY

轻奢风 / LIGHT LUXURIOUS STYLE

FEELING NATURAL PULSE IN THE COURTYARD VILLA
感受自然脉搏的天井别墅

◎ DESIGN CONCEPT 设计理念

The transparent villa of six floors has courtyard green landscape advantage. The designer uses courtyard as central axis, surrounded by space functions such as dendritic extensions, to make natural lights distribute in every corner as if in the sunny church. The space keeps laterally stretched natural scale. Large area of transparent aluminum window frames cut lights into light beams which obliquely enter into the space, which makes residents feel natural temperature all the year round and creates a stack of layering of lights and shadows.

To echo with the leisure and comfortable tone outside the window, the interior uses natural materials to add rich life expression. TV wall made of marbles stretches stairs to outdoor greenery, connecting the interior and exterior fields, which brings bright and capacious effects. The vertical wood elements climb up as if growing plants, manifesting high momentum of the building, which stops at the ceiling and becomes a blooming rose. The elegant smile combines with stamen shape crystal droplight, creating a luxurious field atmosphere.

　　6层楼的透天别墅，拥有天井绿意景观优势，设计师以天井作为中心主轴，周围环绕的空间机能如树枝状延展，让自然光均匀分布在每个角落，如置身在光之教堂。空间内保留横幅开展的自然尺度，通透的大面铝窗框，将光线切割成一道道光束斜射入室，不仅让居住者一年四季都享受到自然温度，也创造出堆栈的光影层次。

　　为呼应窗外休闲舒适的情调，室内运用自然媒材做搭接，增添丰富的生活表情。大理石材打造的电视主墙，延展出通往户外绿意的踏阶，串联起室内外场域，带来明亮宽敞的效果。蔓延至立面的木质元素，彷佛植物生长般攀爬而上，凸显建筑的挑高气势。来到天花处嘎然而止，转瞬成为一朵盛开的玫瑰，绽放优雅的笑靥，结合花蕊状的造型水晶灯，打造奢华的场域氛围。

Project name ｜ Xie Residence
项目名称 ｜ 谢宅

Design company ｜ Yuli Interior Design
设计公司 ｜ 由里室内设计

Designer ｜ Irene Fu
设 计 师 ｜ 傅琼慧

Photographer ｜ Woodman
摄 影 师 ｜ 大头人

Location ｜ Tainan, Taiwan
项目地点 ｜ 台湾台南

Area ｜ 714m^2
项目面积 ｜ 714m^2

Main materials ｜ stone, metal, wood veneer, spray lacquer, glass, bright mirror, tawny glass, etc.
主要材料 ｜ 石材、金属、木皮、喷漆、玻璃、明镜、茶镜等

◎ NATURAL COLORS 自然色彩

Because of the advantage of the building itself, the whole living room is bathed in the embrace of nature. Layers of white ceiling lines present a blossoming rose, collocating with warm yellow custom crystal droplight, which adds a luxurious atmosphere. Two bright yellow chairs in one side of the living room greatly promote fashion flavors of the space. the designer uses pure colors with some blue, yellow and black, which is remarkable.

受建筑本身的优势，将整个客厅沐浴在大自然的怀抱中。层层堆栈的白色天花线条，呈现出玫瑰盛开的意象，搭配暖黄的水晶订制吊灯，增添奢华氛围。而客厅一侧两把亮黄色单椅，大大提升了空间的时尚气息。设计师偏用纯净色系，点睛几笔的蓝调、黄调、黑调，都可圈可点。

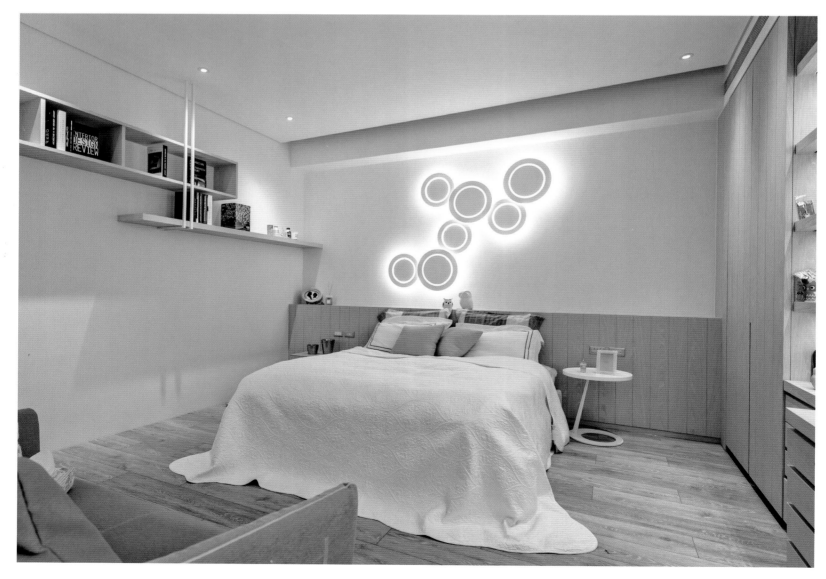

◎ DECORATIVE MATERIALS AND NATURAL LIGHTING 装饰材料、自然采光

The hollow aluminum window frames unify the interior and exterior sceneries into one. Large amounts of natural lights pour into the space, you can feel natural temperature with the change of seasons. Marbles are used to create half tall TV wall in the living room. The display cabinet surface extends to stairs to outdoor greenery, combing with zigzag collage wood veneer facade, which manifests magnificent momentum of the building.

运用铝窗框镂空的造型，让室内和室外景致合而为一，大量自然光倾泄一室，随着四季的变化，感受自然的温度。客厅以大理石打造半高电视墙，透过机柜展示台面的延伸，兼做通往户外绿意的踏阶，结合人字纹拼贴木皮立面，凸显建筑恢宏气势。

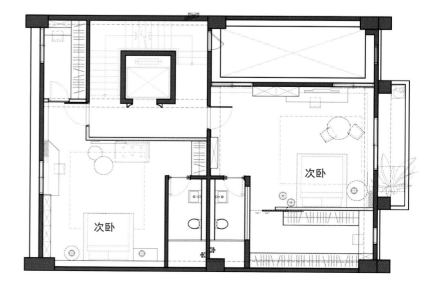

大道至简
SIMPLIFYING PROFOUND PHILOSOPHY

轻奢风 / LIGHT LUXURIOUS STYLE

FASHION SETTLEMENT
时尚聚落

◎ DESIGN CONCEPT 设计理念

Geometric line concept neatly strings this fashion and avant-garde life settlement. Ups and downs, lines are used in the ceiling parting lines in the random, inject modeling images and light paths into the scenes, eliminate the crossing beams and fluently present the tension of the white space. The owner browses previous projects by Jie-yang Interior Design and indicates to use the transparent hanging eye ball chair and angle eye light, which not only brings funny dialogues for the whole layout, but also creates modern and fashion images through design elements which are consistent with the main axis.

几何线条概念，利落串起这场以时尚、以前卫为名的生活聚落，线面起伏、错落之中，串流于天花板的分割线，把造型意象及光线路径，一并编排入目不暇给的场景，也消除了横于空间的梁体，流畅拓延出白色空间张力。而屋主浏览了界阳＆大司室内设计过往设计案，指名使用的透明吊球椅、天使灯，不仅为整体布局带来了点趣味对话，也透过与主轴吻合的设计元素，共铸现代时尚意象。

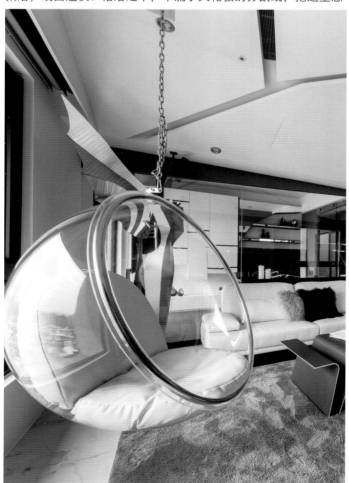

Project name ｜ There is a Kind of Fashion Called Jie-yang
项目名称 ｜ 有一种时尚叫界阳

Design company ｜ Jie-yang Interior Design
设计公司 ｜ 界阳＆大司室内设计

Designer ｜ Ma Chien Kai
设 计 师 ｜ 马健凯

Location ｜ Taipei, Taiwan
项目地点 ｜ 台湾台北

Area ｜ 90m²
项目面积 ｜ 90m²

Main materials ｜ metal, iron part, imported tile, stone, Jie-yang custom-made furniture, etc.
主要材料 ｜ 金属、铁件、进口砖、石材、界阳＆大司订制家具等

◎ DECORATIVE MATERIALS 装饰材料

The craft spirits of software and hardware are the highlights which the designer pays much attention to. Looking at the whole space, the constitutions of custom-made dining table and bar break the existing thoughts of furniture. The unilateral support dining table and extremely thin table top present an asymmetric structure conflict. The other side is embedded into the bar, solving the problem of load-bearing and leading out the bar line with the same sense of design. The jointless artificial surface of 360 degrees and goblet frames with industrial design thoughts create new layout of materials, shapes and structure aesthetics and deduce craft and practical creative concepts.

软硬件的工艺精神，向来是设计师讲究的重点。瞻顾室内，全采订制的餐桌与吧台，其设计构成，就完全突破了既有的家具思维。仅单边支撑的餐桌，以极薄桌板呈现不对称的结构冲突，另一端则嵌入吧台解决了承重压力，同时导引出设计感不相上下的吧台线条。回绕了整整三百六十度，却全无接缝的人造石台面，及上方融入工业设计思考的高脚杯架，皆辟建出材质、形体和结构美感的新格局，演绎工艺与实用集于一体的创造概念。

◎ NATURAL COLORS 自然色彩

The whole space sets black and white as the tone, which manifests black and white tension along with the ups and downs of the space lines. In the sunshine, the black and white space scene presents natural and clear layering. The bright color dining table and chairs strengthen the fashion feeling of the space.

整体空间以黑白定义空间格调，随空间线面起伏、错落，彰显出黑白色彩张力。日光流动下，黑白为主的空间场景，展现出自然、清透的层次。点缀亮色桌椅，加深空间的时尚感受。

大道至简
SIMPLIFYING PROFOUND PHILOSOPHY

轻奢风 / LIGHT LUXURIOUS STYLE

NEW EAST AND HEART LUXURY
新东方 心奢华

◎ DESIGN CONCEPT 设计理念

This space uses montage editing method to direct the exquisite space story.
It relates Oriental cultural memories in different faces. The method is neither a simplification of classical Chinese style nor a simple superposition of Chinese style elements. It digs essences from traditional cultural connotations, keeps traces of Han, Tang, Ming and Qing Dynasties in form, highlights artistic atmosphere by simplified Oriental elements and adds more modern and fashionable elements, which creates a huge difference with the traditional processing technique in the sense of form. The profound and exquisite humanity landscape presents connotations of Oriental temperament and integrates Oriental spirits with luxury closely to reflect distinct flavors collided by hear luxury and new East. Its elegant lifestyle and concise style pursued by scholars manifest a low-key luxury. The combination of humanity and architecture, modern and tradition become a lifestyle from inside to outside.

本空间以蒙太奇剪辑手法来导演细腻的空间故事。

将东方的文化记忆在不同的立面娓娓道出，手法既不是对古典中式的简化，也不是中式元素的简单迭加，而是从传统文化的内涵中挖掘精髓，形式感还保留着汉唐明清的痕迹，以简化了的东方元素突出意境氛围，同时加入更多的现代时尚元素，在形式感上产生与传统处理手法的天壤之别。以更深层次的精细人文山水构图，把东方气质的内涵表现出来，让东方精神与用心营造的奢华密切融合，呈现心奢华与新东方所撞击出的独特韵味。可以见到优雅的生活方式，同时也保存着文人追求简洁的风采，流露出一种低调的奢华，并将人文与建筑、现代与传统融合成为一种从内而外发散的生活态度。

Project name ｜ Crown of the East
项目名称 ｜ 东方之冠

Design company ｜ TINEFUN Interior Planning Ltd
设计公司 ｜ 天坊室内计划

Designer ｜ Qingping Zhang
设 计 师 ｜ 张清平

Location ｜ Taichung, Taiwan
项目地点 ｜ 台湾台中

Area ｜ 2655m²
项目面积 ｜ 2655m²

Main materials ｜ marble, wood, leather, etc.
主要材料 ｜ 大理石、木质、皮革等

◎ DECORATIVE MATERIALS 装饰材料

The floor is paved with whole fine grain marbles, low-key and luxurious, echoing perfectly with the wall marbles. The superior gray rectangle splicing panels echo with the top rectangle gray lens splicing, which is magnificent. What's more, the management of lens enlarges the view of the space and creates good intimate atmosphere.

地面以整面细纹大理石地砖铺成，低调奢华，很好地与墙面大理石材质呼应。而一侧高级灰色调的长方形造型拼接面板，与顶面长方形灰镜拼接相对应，大气磅礴。而且，灰镜的处理，除了扩大了空间的视觉之外，还塑造了良好的私密氛围。

◎ SPACE PLANNING 空间规划

Concise methods are used to handle totem, colors and space lines, which presents changes full of senses of form in low key and restraint and integrates with contemporary style. The whole space layout stresses white space, wishes to leave more possibilities for future and makes the users enjoy life by relaxing attitudes.

简约的手法处理图腾、色彩与空间线条，低调内敛中呈现了富有形式感的变化，与当代设计风格共融互通。整体空间格局强调留白，希望将更多的可能预留给将来，让使用者可以以轻松的态度来享受生活。

大道至简
SIMPLIFYING PROFOUND PHILOSOPHY

轻奢风/ LIGHT LUXURIOUS STYLE

MU-RAY
沐光

◎ DESIGN CONCEPT 设计理念

The owner yarns for quiet and leisure atmosphere of the mountains and expects to endow the space with more comfortable and cozier atmosphere. The concept of "high quality" is the design thought. Starting from life functions, every floor adjusts slightly to gain exclusive and appropriate splendors. Delicate elaboration presents low-key luxury, which ultimately combines humanistic and romantic taste perceptions and emits high quality shiny texture. The three-story villa is located in the quiet mountainous area. The owner does not like excessive explicit luxury, but expects to feel the extraordinary romance in comfortable and cozy atmosphere. Such aspirations and expectations, in fact, are in line with the real meaning "high quality".

屋主向往山林的静谧悠然，更期待空间赋予生活更自在的舒适氛围，以"精品"概念为设计思维，并从生活机能性切入，各层逐步微调出专属、秾纤合度的精彩。细腻铺陈下，更显低调奢华，最终糅合出人文与浪漫兼具的品味感知，散发精品般的光泽质地。共计三层的别墅位于宁静山区，屋主不喜欢过度外显的奢华，但期待在舒适自在中感受不凡浪漫。这样的诉求与期望值，其实正符合"精品"的真正内涵。

Design company ｜ MuzeDesign
设计公司 ｜ 慕泽设计

Designers ｜ Peter Cheng, Sam Tsai
设 计 师 ｜ 郑锦鹏、蔡宗谚

Photographer ｜ Kevin
摄 影 师 ｜ 吴启民

Location ｜ Taipei, Taiwan
项目地点 ｜ 台湾台北

Area ｜ 330m²
项目面积 ｜ 330m²

Main materials ｜ wood grain quartz brick, wood floor, metal, steel brush wood, baked glass, special thin stone, marble, black glass, wood veneer, antique mirror, imported tile, wire frame border, etc.
主要材料 ｜ 木纹石英砖、木地板、金属、钢刷木皮材、烤玻、特殊薄石材、大理石、黑玻、木皮、仿古镜面、进口瓷砖、线板边框等

◎ DECORATIVE MATERIALS 装饰材料

The designers start from the porch to create a magnificent texture. The square and open porch uses marbles to build the wall. The crystal wall lamps and the hollow display cabinet on the left side add beauty to each other. Sitting on the high back chair, you can feel quiet and elegant flavors brought by the corner of the living room. The mountain area has steam so that MuzeDesign team uses a lot of stones. The sofa background wall is covered with marble collages in different shades which creates modern fashion sense. The selection of fresh and clean texture visually is in line with low-key and restraint origin, The wood grain quartz brick and wood veneer ceiling enhance the warmth of the space from up to down. As for the finishing touch of the single product, it is responsible for lightening exquisite atmosphere in the space.

设计师从玄关开始便打造出轩昂的大气质感，方正开阔的玄关，以大理石材为墙体，水晶壁灯与左侧的精品展示镂空柜互相辉映，稍坐在一旁的高背椅处，已然能侧见客厅一隅带来的静谧优雅感受。山区多水气，慕泽设计团队运用大量石材进行铺饰，沙发背墙以不同深浅比例的大理石片拼贴出现代时尚感，挑选视觉上清爽干净的纹理，以符合内敛低调的原则，最后透过木纹石英砖与木皮天花板，上下提升空间暖度，至于画龙点睛的单品，则负责缀亮空间里的精致氛围。

◎ NATURAL COLORS 自然色彩

The third floor is exclusive private space for the couple. The sleeping area and the changing room are on two sides of the stairs. According to the hostess's romantic dream to French atmosphere, the whole floor of the master bedroom area is set with slightly gold noble tone. At the corner of the stairs, black brush gold screen and gold birdcage droplight decorate a French and fancy maturity and romance. In the evening, you can drink a cup of mellow wine, listen to the chanson music to deposit the fatigue of the day, sleep with the stars and look forward to say Bonjour to tomorrow's warm sunshine.

　　三楼是男女主人的专属私领域空间，楼梯两侧分别为睡眠区与更衣室。依照女主人对法式氛围的浪漫梦想，整层主卧区域以略带微金的贵族气息调性铺陈，楼梯终站的转角处，黑色刷金的屏风，金色鸟笼吊灯，精致装饰着法式梦幻的成熟浪漫。夜晚时分，可以选择在此品香醇红酒，听听香颂音乐，借以沉淀一天的疲惫，最后与星辰共枕，并期待与明日的温煦日光说声 Bonjour，日安。

◎ NATURAL LIGHTING 自然采光

In addition to creating a sense of spaciousness, MuzeDesign is also concerned about the amount of mountain lights and designs a space for the owner who loves reading with two benefits, one is good lighting in the dining room, the other is near the dining table and bar there is space for afternoon tea and reading. Based on the functional stratification considerations, two bedrooms and the study are set on the second floor, which also attaches great importance to introduction of light. The study lowers the height of window and designs large areas of window along with the balcony area, which brings in a lot of natural lights and reduces sense of closure by glass sliding door to the outside. In this way, it integrates with the special designed central aisle area into a whole. Whether sitting in a chair of the staircase or walking from the study to balcony, you can feel rich warm sunshine and more humanistic connotations.

除了宽阔感的营造，慕泽设计也在意山区的光线量，并为喜爱阅读的屋主设计了能两者兼具的空间：像是餐厅的采光良好，一旁的餐桌与吧台就能是下午茶或阅读空间；而基于机能分层考虑后，将两间卧室与书房分置于一二楼，也都相当重视光线的导入，书房降低窗户高度后，与一旁的阳台区大面积开窗相呼应，从而导入大量自然光线，而与之对外，更以玻璃拉门来减少封闭感。如此一来，正好与慕泽设计特意打造的宽阔中央过道区域连成一气，无论是坐在梯间单椅上休息，或是由书房走向阳台，都能感染到丰富的阳光暖味，更添人文底蕴。

大道至筒
SIMPLIFYING PROFOUND PHILOSOPHY

轻奢风 / LIGHT LUXURIOUS STYLE

LUXURIOUS CHARM
奢华魅力

◎ DESIGN CONCEPT 设计理念

The house presents distinguished taste of the owner. The designers master advantages of the space, display connotative layering of the field in a unique perspective and inject low-key and luxurious neo-classical vocabularies to meet design blueprint expected by the elites. Opening the door, the living room and dining room of the first floor are presented in a completely open layout full of coherent spacious manner.

居宅样貌呈现屋主的尊贵品味，设计师掌握本案的空间优势，以独特的视角展演场域的内涵层次，并注入低调奢华的新古典语汇，满足层峰人士所希冀的设计蓝图。推开门扉，一楼的客、餐厅以完全开阔的格局呈现，充分感受开放连贯的宽敞气度。

Design company | MuzeDesign
设计公司 | 慕泽设计

Designers | Peter Cheng, Sam Tsai
设 计 师 | 郑锦鹏、蔡宗谚

Location | Hsinchu City, Taiwan
项目地点 | 台湾新竹

Area | 396m²
项目面积 | 396m²

Main materials | marble, crystal steel bake, line board, imported wallpaper, gray lens, tawny glass, imported tile, etc.
主要材料 | 大理石、结晶钢烤、线板、进口壁纸、灰镜、茶镜、进口瓷砖等

◎ NATURAL COLORS AND NATURAL LIGHTING 自然色彩、自然采光

The charm of a better life comes from discovering beauty and possessing beauty. The low-key and luxurious interior collocation comes from love to life. The living room with light color background matches noble blue sofa with gray carpet and curtain, which makes colors of the space dynamic and full of sense of layering and presents echoing harmonious beauty of color. The living room and dining room link together, whose open layout achieves maximum natural lighting. The kitchen sets white stone as the tone. The broad glass window brings in lights and achieves best lighting in the kitchen.

美好生活的魅力来自善于发现美和拥有美，低调奢华内饰搭配源自对于生活的热爱。浅色背景客厅以贵族蓝的沙发搭配灰色地毯和窗帘，使空间色彩跳动富有层次感，又体现出相呼应的和谐色彩之美。客厅、餐厅相连，以开放式的格局实现空间最大化的自然采光效果，厨房以白色石材为主调，宽阔的玻璃窗导入光线，实现厨房最佳采光效果。

◎ DECORATIVE MATERIALS 装饰材料

Entering into the living room, exquisite texture forms a beautiful and exquisite design atmosphere. TV wall uses white marble to present magnificent style. The top adopts layering line board frames, which forms rich transformation of visual sensory. The furniture and lighting configuration choose blue-gray rivets classical sofa to describe beautiful and romantic imagination. The particularly purchased black crystal chandelier of 90 centimeters shines bright light, which is elegance and makes the theme of the space concrete and clear.

进入客厅，以细腻质感构组出精致华美的设计氛围，电视面墙使用白色大理石铺陈恢弘风采，并于上方做层次线板的框塑表现，形成视觉感官的丰富转化；家具与灯光的配置上，选择蓝灰色、铆钉的古典造型沙发，铺述美好的浪漫想像，餐厅则特别购置一盏90公分的黑色水晶吊灯，光芒璀璨、风华绝代，也让空间的主题具象且明确。

DESIGN CONCEPT　设计理念
SPACE PLANNING　空间规划
DECORATIVE MATERIALS　装饰材料
NATURAL LIGHTING　自然采光
NATURAL COLORS　自然色彩
INTERIOR VIRESCENCE　室内绿化
BRINGING SCENERY INTO HOUSE　引景入室

INDUSTRIAL STYLE
工业风

embodying low-key personality, experiencing stress-free rhythm

展现低调随性，体会零压律动

大道至简
SIMPLIFYING PROFOUND PHILOSOPHY

工业风 / INDUSTRIAL STYLE

DESIGN & IMAGINATION, LIVING IN THE HOME OF ONE PIECE
设计 x 想象：住进航海王的家

◎ DESIGN CONCEPT 设计理念

This project consists of five broad and bright floors. The inside space combines creative elements with exquisite craft, creating a dream field.

The receiving area combines with outdoor waterscape, uses penetrable materials such as clear glass and grille and brings natural lights and landscaping into interiors, making the owner see exquisite scenery all around. The second floor uses ocean tone furniture as visual focus. The wall of the living room is covered with steel brush wood veneer with tridimensional world map on it, catching everyone's eyes. The cutting lines restore real longitude and latitude lines, which adds communication and interaction with the space and makes people feel being in the wonderful journey and galloping in the vast global imagination.

本设计由宽阔明亮的五层楼面所构成，空间内部结合了创新元素与精湛工艺，形塑梦想场域。

结合户外水景的一楼迎宾区，使用清玻璃及格栅等穿透感质材，将自然光源及造景带入室内，让屋主环顾四周都是细腻铺述的美景。二楼空间以海洋色调的家俬作为视觉焦点，以钢刷木皮染色的客厅主墙，上方构置世界地图的立体图腾，聚焦所有宾客的目光。切割的线条还原真实的经纬线，除了增加与空间的对话及互动，也让人彷佛身临精彩的旅途航道，驰骋在广阔的环球想象中。

Project name ｜ Fuyu E House Type
项目名称 ｜ 富聿E户

Design company ｜ Yuli Interior Design
设计公司 ｜ 由里室内设计

Designer ｜ Irene Fu
设 计 师 ｜ 傅琼慧

Photographer ｜ Jiarui Zhang
摄 影 师 ｜ 张家瑞

Location ｜ Tainan, Taiwan
项目地点 ｜ 台湾台南

Area ｜ 480m²
项目面积 ｜ 480m²

Main materials ｜ glass, steel brush wood veneer, iron part, marble, etc.
主要材料 ｜ 玻璃、钢刷木皮、铁件、大理石等

◎ DECORATIVE MATERIALS 装饰材料

In order to create an elegant taste and vision penetrable public space, the designer uses ocean tone furniture and lamps as the visual focuses. The dining room and kitchen are equipped with sliding door constituted by iron part and clear glass, which can block out lampblack and is available for friends to party.

营造优雅品味与视野穿透的公共空间，以海洋色调的家俬及设计师款的灯具作为视觉焦点。餐厨区设有铁件及清玻璃构组的拉门，除了阻隔油烟外，亲朋好友聚会时可敞开有效运用。

◎ SPACE PLANNING 空间规划

The bedside wall of the master bedroom uses wave images with the design theme of "travel and ocean" and skillfully covers beam columns of the original structure. What's more, there is an open living area in the master bedroom, which provides a small independent space of the couple. Two droplights like fish lamps are hanged in the dresser, which enriches ocean elements. The cosmetic mirror is removable so that you can enjoy scenery outside the window.

　　主卧床头墙面以波浪的意象，环绕着"旅行、海洋"的设计主题，同时能巧妙地包覆原建物的梁柱。另外主卧房设置了开放式的起居区，让屋主夫妻可以各自拥有独处的小空间。而在化妆台上悬挂着两盏彷佛集鱼灯的吊灯，让海洋元素更丰富，设置的化妆镜面也可左右移动，让人恣意欣赏窗外的景致。

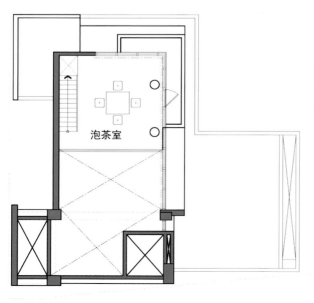

◎ NATURAL COLORS 自然色彩

Bamboo is used in the elder's room, creating a tranquil and sedate atmosphere. The pavement of wood grain tiles makes the leisure tone stronger and is convenient for the owner to clean. The inside natural elements and outside natural tress form a good interaction, which makes the comfortable natural taste stronger.

以竹林意象打造长亲房，营造静谧而沉稳的氛围。同时木纹砖的铺设让休闲调性更强烈，也更方便屋主清理。室内的自然元素与窗外的自然树木形成良好的互动，舒适的自然气息更浓烈。

DANCING SPACE PLAYS A DOUBLE WALTZ
漫舞空间谱奏双人圆舞曲

◎ DESIGN CONCEPT 设计理念

The owner pays much attention to leisure time and is eager to feel complete relax after work. We turn this transparent villa of four floors into performance stage for leisure life and combine the interests of the couple to make unfettered life dance and play a harmonious double waltz.

To show distinct personalities of the couple, the design style develops a new way as if in the New York SOHO district in the 20th century. Without blocks of the partition walls, the spaces connect closely. Large glass window brings in outdoor lights. The "bed BOX" made of wood veneer near the window creates a space effect similar to a stage play through height differences of ground and also creates a natural outdoor atmosphere.

The white and gray space has a large amount of flexible ideas. The facade uses irregular imitated clean water mould materials to present natural plain style, embedded with cement tiles commonly used in early Taiwanese buildings, which presents a conflict beauty of integrating new with old, deduces a personal artistic connotation and willingly creates a living style which belongs to the owner.

相当重视休闲时光的屋主，渴望下班返家后，能感受到全然的放松，我们将4层楼的透天别墅，转化成休闲的生活展演舞台，结合屋主夫妻俩的兴趣喜好，让无拘无束的生活翩翩起舞，谱奏出和谐的双人圆舞曲。

为展现夫妻俩的鲜明个性，设计风格独辟蹊径，宛如20世纪的纽约SOHO区。少了隔间墙阻碍，空间相互重迭紧密，大片玻璃窗让户外光线大量涌入，临窗处木皮构筑出的"卧榻BOX"，透过地坪的高低落差，创造出类似舞台剧的空间效果，也围塑出自然的户外氛围。

白灰配色的空间内，拥有大量灵活创意，立面利用不规则组合的仿清水模材质，呈现出自然的朴质风貌，嵌入早期台湾建筑常见的水泥花砖，展现新旧交融的冲突美感，演绎自成一格的艺术涵养，随心所欲创造出属于屋主的居家风格。

Project name ｜ Liao Residence
项目名称 ｜ 廖宅

Design company ｜ MU SPACE DESIGN
设计公司 ｜ 慕森设计

Photographer ｜ WEIMAX STUDIO
摄 影 师 ｜ WEIMAX STUDIO

Location ｜ Taichung City, Taiwan
项目地点 ｜ 台湾台中

Area ｜ 175m²
项目面积 ｜ 175m²

Main materials ｜ custom-made iron part, glass, imitated clean water mould paint, steel brush wood veneer, blackboard paint, archaistic brick, etc.
主要材料 ｜ 订制铁件、玻璃、仿清水模漆、钢刷木皮、黑板漆、仿古石砖等

◎ DECORATIVE MATERIALS 装饰材料

The floor of the porch is covered with imitated metal bricks to present the original surface. The shoe cabinet door piece uses ceramic paint baking processing technique to bring delicate fog texture. The iron part and glass define the dust area. The back of the shoe chair uses blackboard paint and metallic paint to create a unique and creative message graffiti wall. Opening the door with bright yellow iron part handles, you can enter into the dance room in the first floor which becomes dancing space for the couple to spend their spare time.

玄关地坪铺设仿金属砖呈现原始表面，鞋柜门片采用陶瓷烤漆加工法，带来精致的雾面质感。利用铁件玻璃界定出落尘区，穿鞋椅背后以黑板漆和磁性漆，打造出独具创意的留言涂鸦板墙。推开采用跳色处理的鲜黄色铁件门把，即进入到一楼的舞蹈室，闲暇之余成为男女主人的漫舞空间。

◎ SPACE PLANNING 空间规划

The second floor is public area in open design, which links living room and dining room closely and brings natural lights into interior. The bed area makes use of small spaces caused by beams to plan a rectangular recreational couch to lighten the oppressing sense brought by beam lines. At the same time the walls and floors are covered with cedar wallboards, collocating with natural lights, which creates a natural and leisure atmosphere.

二楼空间以开放式设计规划为住家公领域，不仅客厅和餐厅紧密相连，也让自然光走入空间内部。卧榻区则善用横梁产生的畸零地，规划口字型的休闲卧榻，淡化梁线带来的压迫感。同时以温润的香杉壁板铺陈天地壁，搭配黑色百叶窗筛落自然光束，打造自然休闲的氛围。

◎ NATURAL COLORS 自然色彩

The dining room uses white as the basement, collocating with gray paint, which extends lateral scale of the space, in addition with iron part display cabinet and old dining table and chairs, bringing conflict visual aesthetics. The master bedroom uses blue iron part door frames to bring clear field definition, collocating with bright yellow tea table and soft ware, which jumps out of the gray and white space. What's more, the functional area of the master bedroom combines bookcase with collecting closet, creating retro and multi-functional designs. The open closet uses sliding door to create flexible screening effects, collocating with tube modeling iron part to add line rhythm to the cabinet, which emits warm loft style.

餐厅以白色为基底搭配灰色漆面，延展出空间的横幅尺度；结合铁件展示柜及仿旧餐桌椅的摆放，带来冲突的视觉美感。主卧以蓝色铁件造型门框带来明确的场域界定，搭配鲜黄色边几和软件，跳脱出灰白色的空间铺陈。此外主卧的功能室以书柜与收纳衣柜做结合，打造出复合多功能设计。开放式衣柜利用谷仓滑门创造弹性的遮蔽效果，搭配亚管造型铁件增加书柜的线条律动感，散发温暖的Loft风貌。

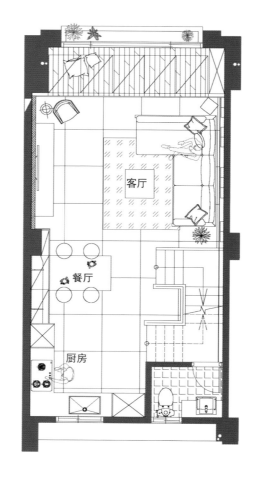
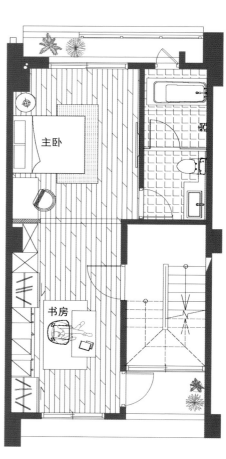

ACTIVE FACTORS HIDDEN IN THE ZEN-LIKE TONE
藏在禅风基调里的活泼因子

◎ DESIGN CONCEPT 设计理念

The living space of three generations needs to satisfy the living conditions of four bedrooms and two living rooms and has sense of space. the designer slightly adjusts the layout to endow this project with complete life functions, embeds niche for a statue of the Buddha expected by the elders in the corner and uses the integration of the entire style to combine aesthetics with functions to create a texture and perfect residence for the family.

三代同堂的居家空间，需满足四房两厅的起居条件，且同时要具备空间感，设计师透过格局微调的方式，赋予本案完整的生活机能，更利用畸零角落嵌入长辈期望的佛龛，再借由整体风格的整合，将美感与机能相互融合，替屋主一家人创造质感满分的完美居宅。

Design company ｜ Dayoungdi Design
设计公司｜大漾帝国际室内装修有限公司

Designer ｜ TAI MING CHUAN
设 计 师｜戴铭泉

Photographer ｜ Hey!Cheese
摄 影 师｜Hey!Cheese

Location ｜ New Taipei City, Taiwan
项目地点｜台湾新北

Area ｜ 130m²
项目面积｜130m²

Main materials ｜ natural wood veneer, marble, iron part, special paint, rock slice, cloth, etc.
主要材料｜天然木皮、大理石、铁件、特殊涂装、岩片、织布等

◎ DECORATIVE MATERIALS 装饰材料

The corridors to privates areas are paved with warm wood which hides the bedroom door pieces, creating a whole and consistent visual beauty. Below is equipped with footlights to illuminate at night. What is called "the devils hide in details", in addition to space aesthetics and function arrangements, the designer stresses details of the project. The well-chosen building materials, exquisite methods and solid hardwire furnishings promote the integrity and durability of the living space and give residents life quality of living for a long time without worries.

　　通往私领域的廊道利用温润木质做铺陈，并将卧室门片隐藏其中，创造整体一致的视觉美感，下方更贴心嵌入踢脚灯，提供夜间照明需求。正所谓"魔鬼藏在细节里"，除了空间美感与机能安排，设计师同时强调工程的细节，透过精心严选的建材、精密的工法及坚固的五金配件等，提升居家空间的完整度及耐用度，才能给予居住者久居无忧的生活质量。

◎ SPACE PLANNING 空间规划

In order to gain broader view after entering inside, the designer firstly dismantles wall partition of the kitchen to make the living room and dining room and kitchen form an open kinetonema, then moves the TV wall back to leave more activity space, which enlarges public space and makes people feel the broad view. In the visual warm wood, put off the tired body and mind, you can gain strengthens to calm your mind.

为了让入门后的视野更加开阔，设计师首先拆除厨房的实墙隔间，让客厅与餐厨区形成开放式动线，并将电视主墙向后挪移，替客厅争取更多活动空间，进而放大公领域尺度，让人一进门就感受到辽阔视野，在一片木质温润的视觉铺陈中，卸下一日疲惫的身心，获得安定心绪的力量。

◎ NATURAL COLORS 自然色彩

The interior sets the owner's favorite Zen style as the tone. A lot of wood colors are used in the ceiling and walls, bringing a warm texture to the space. Concise white and field temperature are partly used to avoid sedate wood elements of being depressive. Bright lights are leaded in to illuminate the interior, creating natural and bright leisure atmosphere. In addition, in order to make the space jump out of the rich expression changes, the ceiling of the living room adopts oblique paving method to lead visual rhythm. The sofa background wall is paved with green bronze finish, creating a mottled and old facade effect. Granites with fresh grains are used in the TV wall, creating rough and magnificent field momentum.

室内以屋主喜爱的禅风为定调，大量的木色语汇游走天花及壁面，带来一室温暖质地，局部带入简约的白适度调和场域温度，避免沉稳的木质元素让空间显得压抑，且引进室外饱满采光照亮室内，衬托自然明亮的休闲氛围。此外，为了让空间跳脱出丰富的表情变化，客厅天花板特别采用斜向铺排的方式，引领视觉的律动性；沙发背墙透过青古铜面漆，铺陈出斑驳仿旧的立面效果，以及选用纹路鲜明的花岗石作为电视主墙，营造粗犷大气的场域气势。

工业风 / INDUSTRIAL STYLE

大道至简
SIMPLIFYING PROFOUND PHILOSOPHY

WARM GRAY AESTHETIC SITUATION
暖灰的美学情境

◎ DESIGN CONCEPT 设计理念

The magnificent green and mountain ridge outside the window is the key to the concept generation. The tooled finish rock slab embedding the lofty and steep mountains and the iron accessories demonstrating the continuous wall line, combining the definition of smooth generatrix and open field, are designed in order to introduce the beauty of the majestic and peaceful mountains as well as provide user a customized public sphere with a theme of "gathering friendship".

In addition, all of the design originality in the house constitutes a complete composition with starting and end point, which brings out the aesthetic and practical design taste. In particular, the parallel functional axis in the kitchen and dining area, together with the creative TV wall taking advantage of thinking transposition produce a whole new life experience. The original pattern with three rooms is transformed into two master bedroom and a leisure study with fluent and free life generatrix. The integration of material details, design style and collecting functions create enjoyable and pleasant resident for the couple who is engaged in the top of advertising and marketing industry.

窗外壮丽的苍翠棱线，一开始就是发想关键，为了将山的雄浑与安定之美引入空间，特地以凿面峥嵘的岩板，搭配铁件演绎连续墙线，并结合流畅动线与开放式场域定义，为使用者量身客制以"聚·交谊"为题的公领域。

此外，宅中所有巧思都是有起点、有终点的完整构图，由此激荡出兼具美感与实用的设计品味，而餐厨区并列的机能轴线，加上换位思考的电视墙创意，更为生活带来全新体验。将原有3房的格局，改为双主卧加休闲书房的形式，并规划出流畅自如的生活动线，给予材质细节、设计风格、收纳机能的融会贯通，为从事广告营销业的金字塔顶端夫妻俩，打造享乐生活的自得宅邸。

Project name | Extended Mountain View
项目名称 | 衔岩

Design company | YYDG INTERIOR DESIGN
设计公司 | 源原设计

Designers | Peny Hsieh, Calvin Tsai
设 计 师 | 谢佩娟、蔡智勇

Location | Taipei, Taiwan
项目地点 | 台湾台北

Area | 150m²
项目面积 | 150m²

Main materials | stone skin, waterproof paint, imitated rust paint, spray lacquer, oak, stone, etc.
主要材料 | 石皮、水磨漆、仿锈铁漆、喷漆、橡木、石木等

◎ DECORATIVE MATERIALS 装饰材料

The wall of the porch uses warm gray and wood to create an extensive axis to the residence. The left wood texture spreads to the public area to enlarge it and meet the needs of reception and gathering, collocating with window scenery and lighting, which endows the space with open sense. Sofa background adopts a whole rock plate embedded with iron part layer board to deduce natural expression of the facade, collocating with texture furniture, which presents texture and modern life atmosphere. TV wall is set near the dining room and kitchen and covered with imitated rust paint, collocating with the kitchen ware configuration, which forms magnificent visual wall, at the same time collocating with a long solid wood table, which brings innovative audio-visual and dining experience.

　　玄关墙面由暖灰与木质对话，塑造进入居宅的延伸轴线。左侧的木头肌理，转折延伸到公共厅区，以放大为诉求，满足接待及团聚需求，搭配窗景延揽绿意采光，赋予空间无比的开阔感。厅区沙发由一整面岩板嵌上铁件层板，演绎立面的自然表情，搭衬质感家具的铺排，演出质感都会的生活气氛。电视主墙则安置于餐厨区旁，以仿锈铁特殊漆铺陈表面，并连结厨具设配，形成大气磅礴的视觉主墙，同时搭配实木长桌延伸，带来新颖的视听与用餐感受。

◎ BRINGING SCENERY INTO HOUSE AND NATURAL COLORS　引景入室、自然色彩

Pursuing life aesthetics, the wall uses natural materials and earth tone to bring visual layering. Warm wood color of the porch leads returned people fall into home's arms without any hesitation. Black rock wall of the living room and laterally arranged iron part framework include life collections and deconstruct the most touching scene, collocating with longitudinally arranged sofa and chairs with neat lines, which forms a rough and delicate contrast and shape comfortable space personality initially given to the casual living room. In addition, there is a reading and leisure area behind the sofa, with some black and gray chairs, which is as if custom-made for the space without any sense of interruption.

讲求生活美学呈现，壁面自然材质和大地色系挥洒，带出视觉层次感。入门玄关温暖的木色，引领归者毫不犹豫投入家的怀抱。客厅黑色岩墙与横向错落的铁件格架，包容生活上的搜藏，解构出最动人的框景，再搭配纵向排列的沙发与线条利落的单椅，粗犷与细腻的对比，雕塑最初赋予随性客厅、舒适的空间性格。此外，沙发背后特别设计出一块阅读休闲领域，摆上披着黑与灰外衣的单椅，彷佛为这空间量身定制，毫无违和感。

◎ SPACE PLANNING 空间规划

The black lamp groove of the ceiling initially parallels with the black rock wall of the living room, after twice turns develops downward and connects with the floor along the wood modeling wall, which acts as lighting, leads out the main wall vision and sorts out consistency of the space. The designers fully master features of natural materials, divide fluent and free layout generatrix and aesthetic practice and redefines pleasure life field for the couple.

天花的黑色灯沟一开始平行客厅黑色岩墙，两度转折后向下发展，沿着木色造型墙衔接地面，不只发挥照明功能，更引导出主墙视觉，整理出空间的连贯性。设计师在充分掌握天然材质的特色下，划分流畅自如的格局动线与美学实践，重新定义了夫妻俩的享乐生活场域。

大道至简 | SIMPLIFYING PROFOUND PHILOSOPHY

工业风 / INDUSTRIAL STYLE

TIMELESS
慢行

◎ DESIGN CONCEPT 设计理念

Located at Tamsui City overlooking the estuary of Tamsui River under Mount Guanyin and featuring a private home spa pool in each unit, this famous luxury residential tower manifests a housing concept comprising pool views, river views and sea views. To extend into the interior magnificent scenes of colorful water ripple reflections during Tamsui sunset, a creative design approach has been adopted with Italian Memento antiqued-effect porcelain tiles paved on the floor in great areas interpreting genus loci, a pervading spirit of the place, while handmade faux stone uneven finish echoes the ruffling tiny waves on water surface. With magical power exerted by time, all activities and moods in this vacation residence seem to be slowing down. Things though sharing the same space and time, illusion of time dilation appears.

本案位于淡水的名宅特色是一户一泳池，远眺观音山淡水河出海口，在表现这样池景河景甚至海景的住宅精神下，我们希望将淡水夕照下水面波光粼粼的印象延续至室内场景，因此空间中主要大面积呈现场域精神的地坪以意大利memento仿旧复古砖，手工不平整的仿石纹立体面造就其波光明净的效果。作为一个度假住宅，时间也是相当有魔力，一切的行为与心境皆慢了，虽然同样运行于一个时空，却产生如时间膨胀的错觉。

Design company ｜ Waterfrom Design
设计公司 ｜ 水相设计

Designers ｜ Nic Lee, Karen Lin, Richard Kuo
设 计 师 ｜ 李智翔、林怡慧、郭瑞文

Photographer ｜ Sam Tsen
摄 影 师 ｜ 岑修贤

Location ｜ Taoyuan, Taiwan
项目地点 ｜ 台湾桃园

Area ｜ 462m^2
项目面积 ｜ 462m^2

Main materials ｜ travertine, granite, hand-made brick, hand-made paint, leather, wallpaper, etc.
主要材料 ｜ 洞石、花岗石、手工砖、手工漆、皮革、壁纸等

◎ NATURAL LIGHTING 自然采光

We want to produce a condition opening to the outside world in every direction in the closed interior compared with the open river view and create the effect of extending to the boundless surrounding space. Therefore all the behavior patterns basically face with the Tamsui Rive including the cooking and sports. All the scales of the corridor are open and the kinetonema of the space is rounding and circulatory. There is no furniture leaning against the wall to block the flow of the space. The sofa, island kitchen and gym can be saw in every view without limitation. The shoe cabinet at the porch entrance abandons large area in constructivism, separates four cabinets by structure and combination and guides lights into the central space emotionally and rhythmically.

我们希望在相对于开放河景的封闭室内，能产生一种各面都向外界开启的状态，能延伸至周围无限空间的效果。因此所有的行为模式基本都是面向淡水河，包括下厨烹饪以及运动健身；所有走道的尺度都是放宽的，空间的动线都是环绕循环，没有靠在墙的家具阻碍空间的流动脉络，包括沙发、中岛厨房甚至健身房都能在各面向角度观看不受限制。玄关入口的鞋柜以构成主义舍弃大面积的量体，以组构和结合分离四座柜体，将光有情绪节奏地导入空间中央。

◎ DECORATIVE MATERIALS AND NATURAL COLORS 装饰材料、自然色彩

A distinctive atmosphere of frozen timelessness in the place is what we aspire to catch with the flowing river in front and the passing sun overhead left as the only moving things. Neutral colors and earth tones, such as beige in the leather main wall, beige grey in floorings, light brown green granite back wall and off-white in travertine main wall have been used to convey moods of placidity and steadiness of the space, while the shared feel and warmth of these grained materials as well as the perpetualness communicated by their being simple and unadorned respond to Piet Mondrian's usage of primary colors. Colors help set boundaries to a space rather than decorate it.

我们希望这个场域的空间精神像是冻结住的慢行，只剩下前方的河水在流动，只留下顶上的太阳在游移。因此空间中利用中性的色彩，如米色系的皮革主墙、米灰色的地坪、浅棕绿色系的花岗石背墙与米白洞石主墙等大地色系，让空间画面传递平和稳定的情绪，而这些材质的共通性强调纹理的触感与温度，朴实无华透过材质传递永恒的时间性，也一如蒙德里安只采用原色，色彩的功能不在于装饰，而是辅助空间的界线。

大道至简
SIMPLIFYING PROFOUND PHILOSOPHY

工业风 / INDUSTRIAL STYLE

VIVID COLOR
重彩

◎ DESIGN CONCEPT 设计理念

Inspired with paper patterns created by fashion designers for garment making to interpret their imagination, this interior design, taking silhouettes from such patterns, starts dialogues with the young residence owner, a French educated female fashion designer. Just as dress drawings, folding and twisting in pattern making are transformed to clean-cut presentation of sharply defined copper tubing of pendant lamps hanging down from ceiling and wires braided on the glass bookcase.

衣物纸样是服装设计师想象空间的陈述、制衣的草本，为呼应年轻女业主留法服装设计师之身份，汲取纸样线条而成就设计灵感。打板线描下的折迭扭曲，在书柜与天花转化为铜线的利落排列，一如服装图纸。

Design company ｜ Waterfrom Design
设 计 公 司 ｜ 水相设计

Designers ｜ Nic Lee, Siting Lv, Xi Chen
设 计 师 ｜ 李智翔、吕思亭、陈皙

Photographer ｜ Lee Kuo-Min
摄 影 师 ｜ 李国民

Location ｜ Taipei, Taiwan
项目地点 ｜ 台湾台北

Area ｜ 76m^2
项目面积 ｜ 76m^2

Main materials ｜ metal iron casting, metallic mesh, stone brick, valchromat, mirror, glass, veneer, etc.
主要材料 ｜ 金属铁件、金属网、石材砖、沃克板、镜面、细纹玻璃、木皮等

◎ DECORATIVE MATERIALS 装饰材料

Fabrics are what it takes to make garments. Prompted by silks and lace, the translucent texture is reinvented as metallic mesh and glass in the bookcase to reflect see-through effects as well as the contrast between the glossy glass and the unadorned slate wall creating an artistic taste with converted optical illusion showcasing the owner's keen interest in reading fashion books. With the figurative banking out to the minimal abstract design elements of the bookcase incorporate forms and shapes of bow ties and belts from garment accessories and Art Deco is exhibited as delicate details within bookcase interior.

制衣的媒材是衣料，汲取蕾丝、绫罗布料之纹理、透视，相映书柜金属 mesh 网格、玻璃镜面之透视实虚，延伸出玻璃与板岩砖墙的相对，创造效果再转化的别趣错视，扶引业主喜翻阅浩繁服饰藏书的习惯。如衣物之于饰品，柜体设计元素涵纳领结、皮带等服饰形体，将具象消隐至抽象极简，Art Deco 成为柜体与空间的精致细节。

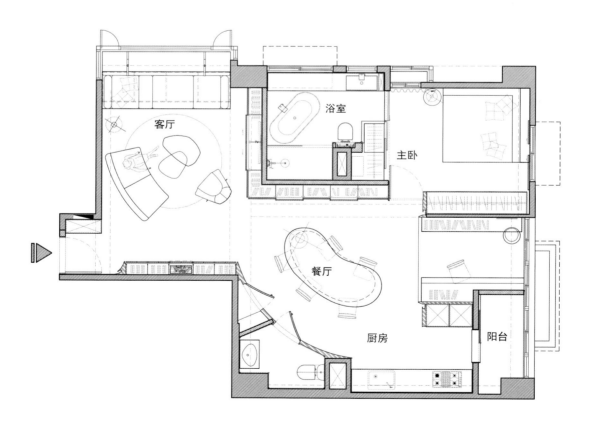

◎ NATURAL COLORS 自然色彩

Inspired by artworks of René Gruau, a well-known fashion illustrator, colors employed in the space take on the look of a French-style vividness and audacity. The study in warm orange, the living room in tranquil blue and the kitchen in sumptuous gold complement and contrast with one another. The thick yet bright colors in large areas bring in a taste of rhythmic and geometric montage collocating the elegant charm of linear adornments. With a look suggesting a sewing machine the tailor-made base of the dining table breaks through hedging-in traditions of balance with asymmetric fashion deconstructing and echoes Experimentalism in the fashion design trend.

灵感撷取时尚插画大师 René Gruau 之作，反映着法式用彩之活泼大胆，成为空间色调的摘取。温暖橘色书房、静谧蓝客厅与华贵金色厨房彼此相搭相应，以鲜颜重彩、大面积体的色块律动及几何拼接趣味，与线性装饰搭配起优雅韵味。如不对称的时尚解构，餐桌桌脚订制如缝纫机具，打破经典平衡的冲突，相衬服装设计潮流的实验派性质。

大道至简
SIMPLIFYING PROFOUND PHILOSOPHY

工业风 / INDUSTRIAL STYLE

MEMORY, DIMENSION DOOR
记忆，任意门

◎ DESIGN CONCEPT 设计理念

Originated from that home is the space which carries memories and with the concept of "dimension door of memory", we think door is not only the entrance of spatial transformation but also the key to open memories so that the door is designed as the image of home. There is an unknown world behind every door as if the dimension door through which you can freely shuttle between times and spaces endows every room with exclusive life memory.

The theme wall spreads widely in the dining room and extends natural rhythm as the first side view after entering inside. The decorative wall boldly shows the red bricks. Iron part display cabinet brings geometric visual effects and creates tensions of the space. The sideboard with line plate elements is painted with old brush color, which is full of historic traces and French classical rural amorous feelings. Two mottled doors as if opening the door of memory make the space narrate stories with memories. The vision of the kitchen is open and clear and the bright tone amplifies sense of space. The ceiling is decorated with warm wood veneer, which jumps out of the shallow space and creates a warm life field.

缘起于家是承载记忆的空间，以"记忆的任意门"为概念，我们认为门不只是空间转换的口径，也是开启记忆的钥匙，便将门形塑成家的意象。每扇门后面藏着未知的世界，就像可自由穿梭空间的任意门道具一样，让每个房间内都拥有独家的生活记忆。

餐厅区宽幅展开的主题墙柜，蔓延自然的律动感，作为入室后的第一道端景。大胆裸露出红砖砌成的装饰墙面，以铁件展示柜带来几何造型的视觉效果，创造出十足的空间张力。餐边柜加入线板元素，经仿旧刷色处理，充满岁月历史痕迹，颇富法式古典乡村风情。巧妙纳入两扇斑驳造型的门片，仿佛推开一道记忆之门，让空间开始诉说起有记忆的故事。来到厨房视觉豁然开朗，借由明亮色调提升空间放大感，天花板装饰温润的木皮，跳脱出轻浅的空间铺陈，打造充满温度的生活场域。

Project name ｜ Qiu Residence
项目名称｜邱宅

Design company ｜ MU SPACE DESIGN
设计公司｜慕森设计

Photographer ｜ WEIMAX STUDIO
摄 影 师 ｜ WEIMAX STUDIO

Location ｜ Taichung City, Taiwan
项目地点｜台湾台中

Area ｜ 166m²
项目面积｜166m²

Main materials ｜ custom-made iron part, glass, imported tile, steel brush wood veneer, art lacquer, blackboard, soil, etc.
主要材料｜订制铁件、玻璃、进口花砖、钢刷木皮、艺术漆、黑板、乐土等

◎ DECORATIVE MATERIALS 装饰材料

The old cabinet door piece and soil facade in the porch presents concise and neat gray tone. The Italian retro tile floor jumps out of the space layering. TV wall is covered with environmental protected soil and the below is embedded with iron part, which manifests rough sense of space. Hand-made white concrete wall of sofa background presents hand-made grain layering and brings unique surface texture. The dining room uses soil and wood floor to define the field. The sofa background wall uses part of lattice modeling to extend the sight.

玄关仿旧柜体门片和乐土立面，呈现灰系的简约利落，以意大利复古花砖地坪，跳脱出空间层次。电视墙以环保乐土塑造表面，下方内凹空间镶嵌铁件，展现粗犷的量体感。沙发背墙手刮成形的白色水泥墙，展现手作的纹理层次，带来独特的表面触感。餐厅以乐土和架高木地板，带来场域分野界定；沙发背墙采局部格窗造型，让视线得以延伸。

◎ NATURAL COLORS 自然色彩

The master bedroom uses many kinds of gray culture stone, which creates a progressive effect of gray stairs under natural lights. The master bathroom uses colorful Spanish ceramic tiles, which creates outstanding designs and endows the showing area with wonderful views. The subaltern room uses spaces under beams to set down the lake blue cabinet, collocating with the nearby bed, which shows beauty with collecting function.

主卧以多种灰色文化石堆砌，在自然光的照拂下，创造出递层的灰阶效果。主卫则以色彩丰富的西班牙陶瓷花砖，创造出亮眼的设计，让淋浴区拥有纷呈视觉。而次卧善用梁下空间规划湖蓝跳色柜体，搭配一旁的卧榻，具有展示美观和收纳功能。

图书在版编目（ＣＩＰ）数据

大道至简：台式空间设计品鉴 / 深圳视界文化传播有限公司编. -- 北京：中国林业出版社，2017.3
ISBN 978-7-5038-8919-6

Ⅰ．①大… Ⅱ．①深… Ⅲ．①室内装饰设计－作品集－台湾－现代 Ⅳ．① TU238.2

中国版本图书馆CIP数据核字（2017）第 023416 号

编委会成员名单
策划制作：深圳视界文化传播有限公司（www.dvip-sz.com）
总 策 划：万绍东
编　　辑：杨珍琼
装帧设计：黄爱莹
联系电话：0755-82834960

中国林业出版社 · 建筑分社
策　　划：纪　亮
责任编辑：纪　亮　王思源

出版：中国林业出版社
（100009 北京西城区德内大街刘海胡同 7 号）
http://lycb.forestry.gov.cn/
电话：（010）8314 3518
发行：中国林业出版社
印刷：深圳市雅仕达印务有限公司
版次：2017 年 3 月第 1 版
印次：2017 年 3 月第 1 次
开本：235mm×335mm，1/16
印张：20
字数：300 千字
定价：398.00 元（USD 79.00）